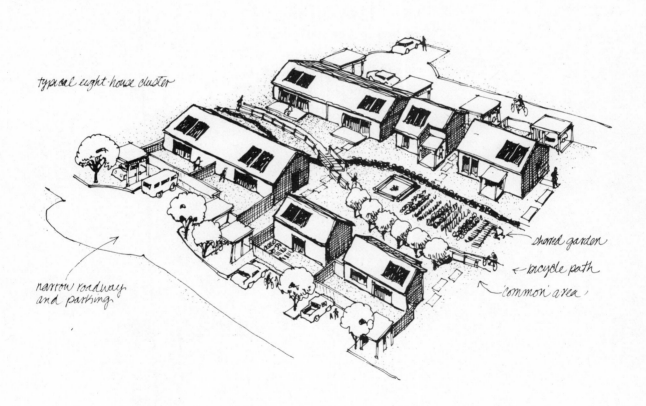

typical eight-house cluster

narrow roadway and parking

shared garden

bicycle path

common area

PRESENT VALUE

Yolo Environmental Resource Center

Friends of the Earth
Principal offices

124 Spear Street
San Francisco, California 94105

530 Seventh Street, S.E.
Washington, D.C. 20003

72 Jane Street
New York, New York 10014

PRESENT VALUE
Constructing a Sustainable Future

Gigi Coe
Office of Appropriate Technology
State of California

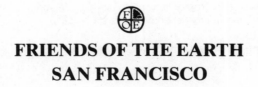

FRIENDS OF THE EARTH
SAN FRANCISCO

Library of Congress Catalog Card Number:
79-90491

ISBN: 0-913890-35-9

Cover design by Hal Lockwood
Designed and edited by Nancy Austin

Trade sales and distribution by Friends of the
Earth, 124 Spear Street, San Francisco, California
94105

The Office of Appropriate Technology (OAT) as-
sists and advises the Governor and all state agencies in
developing and implementing less costly and less ener-
gy-intensive technologies of recycling, waste disposal,
transportation, agriculture, building design, and resource conservation. It was established by Governor
Brown in May of 1976. Our staff is divided into four
activity areas: design and technical assistance, community assistance, publications and information, and energy programs. Bob Judd is the Director of OAT, which
is a part of the Governor's Office of Planning and Research.

OAT welcomes comments and questions about
its programs. Write to OAT, 1530 Tenth Street, Sacramento, California 95814, for a list of available publications.

Project staff: Research, writing, production: Gigi
Coe. Design, illustration, photography: Bill Wells, San
Anselmo. Technical consultants: Scott Matthews, Ken
Smith, Dave Rozell, Mack Walker, and Tyrone Cash-
man. Editor: Hal Rubin. Research assistance: Susie
Cahill, Mary Lou Van Deventer, and Judy Michalowski.
All photographs by Bill Wells unless otherwise noted.

Our thanks to Ann Bartz, Allan Lind, and Jerry Yudelson,
who helped improve the book by reviewing the manus-
cript. Thanks also to Janet Bandy, Lily Satow, and Che-
ryl Yee, without whose patient and skillful support the
project couldn't have proceeded.
Our special thanks, too, to all the people whose
work appears in this book, to Bob Judd, Kirk Marc-
kwald, and Scott Matthews for their continuing contributions to this project, and particularly to Wilson Clark for
his ideas, encouragement, and participation.

Printed in the United States of America

PRESENT VALUE
Constructing a Sustainable Future

From solar-heated homes to community gardens, this book illustrates efforts that have been made in California to build for a future rather than steal from it.

In the briefest period, we are witnessing a shift from the Industrial Age to the Solar Age. The effect will be felt in our homes, our businesses, our families, even in our patterns of thought. From waste and obsolescence we will awaken to care and conservation.

Present Value can be sustained by choosing wisely and taking into account what we now know about life-cycle costs versus the illusory advantages of lowest first cost. This book provides some examples of specific conservation efforts. There are many others. The rest is up to you.

Sincerely,

Edmund G. Brown, Jr.
Governor,
State of California

CONTENTS

FOREWORD

Many recent visitors to rural areas in the Third World have encountered small biogas plants—locally constructed tanks that convert animal dung into methane gas and fertilizer. The methane produced in these plants is a clean, efficient fuel for cooking and heating, and the high-quality fertilizer is free of the pathogens originally in its ingredients. Some 4,300,000 biogas plants have reportedly been built in China in the last four years, many of them large, communal affairs. In Korea, Pakistan, Kenya, and Costa Rica, the devices blend so well into local surroundings that few visitors are aware of the shadow of a white-haired Englishman by the name of E.F. Schumacher looking behind them.

Biogas plants are examples of what development specialists call "appropriate technologies." The meaning of this term can be illuminated by contrasting an appropriate technology with an "inappropriate" one—inappropriateness is judged by local conditions.

In India today there is a crying need for nitrogen fertilizer. A large coal-fueled fertilizer factory will produce 230,000 tons per year; 26,000 biogas plants are needed to produce the same amount. But the biogas plants would cost $15 million less to build, and all this money would be spent inside India, saving $70 million in foreign exchange. The biogas plants would provide 130 times as many jobs, and these jobs would be in the rural villages where most people live and where employment is most desperately needed. Biogas plants would also produce the fertilizer where it is needed, eliminating the transportation required from the centralized coal plant. Finally, the biogas plants would produce enough fuel each year to meet most of the energy needs of 26,000 Indian villages, while the coal-fired plant would consume enough fuel every year to meet the energy needs of 550 villages.

Biogas plants, in short, meet the criteria set forth in the late E.F. Schumacher's most important book, *Small is Beautiful:* "Technology of production by the masses, making use of the best of modern knowledge and experience, conducive to decentralization, compatible with laws of ecology, gentle in its use of scarce resources, and designed to serve the human person instead of making him the servant of machines."

In brief, appropriate technology can offer us the basis for a sustainable civilization: one offering equity, social justice, and human dignity—not only for developing countries, but for the industrial world as well. In the retrospective vision of our grandchildren, sustainability and equity may prove far more important than the narrower, shorter-sighted criteria that dominate current decision making.

Perhaps the most intriguing aspect of the A.T. revolution lies in its social and political ramifications—factors not considered in most policy analyses. Energy decisions are now based on the naive assumption that competing energy sources are neutral and interchangeable, for example. As defined by most energy experts, the task at hand is simply to obtain enough energy to meet projected demands as inexpensively as possible; cost is the deciding factor.

The error is that all energy sources are not alike; they are neither neutral nor interchangeable. Each possesses a unique, inherent set of characteristics—some good, some bad. Some energy sources are necessarily centralized; others are necessarily dispersed. Some are exceedingly vulnerable; others are nearly impossible to disrupt. Some will produce many new jobs; others will reduce the number employed. Some will tend to diminish the gap between rich and poor; others will accentuate it. Some inherently dangerous energy sources can flourish only under authoritarian regimes; others can lead to nothing more dangerous than a leaky roof. Some sources can be comprehended only by the world's most elite technicians; others can be assembled in remote villages using local labor and indigenous resources.

Around the globe, people, communities, and countries are committing themselves to appropriate technologies, discovering that they can use native talent, resources, and financing to solve their own problems. And, they can do so today.

Brazil, far more dependent on OPEC than the U.S. is, has committed itself to severing that oil bond by the year 2000. The country is well on its way toward an ethanol-based economy, produced from sugarcane and cassava, which are abundant. By the turn of the century, ethanol is intended to replace all gasoline in Brazil.

A tenants' organization in New York City has discovered that appropriate technology works in the heart of the city. In a formerly broken-down neighborhood on the Lower East Side, the tenants of 519 East 11th Street acquired a burned-out building through a sweat-equity program. They renovated the building, insulated the walls, and installed solar water heaters and a reconditioned wind generator. Forming an organization called the Energy Task Force, they now spread their knowledge through publications and workshops for people with similar interests.

Appropriate technology encompasses far more than just energy production. Frugality and reuse in all areas can create tremendous savings in capital, energy use, and environmental impact. Most of the materials Americans now throw away could be reused without significant lifestyle changes. With products designed for durability and recycling ease, the waste streams of the industrial world could be reduced to a small trickle. Recycling and reuse can dramatically reduce energy demand. Aluminum is a superb example of such savings. Production of one pound of aluminum from virgin ore requires 134,000 BTU's; production from recycled materials requires just 5,000 BTU's. If steel mills and foundries reused ferrous scrap, it would reduce air pollution 86%, water pollution 76%, and water use 40%, and virtually eliminate solid waste.

A classic success story demonstrating the interdependence of capital savings, materials reuse, energy conservation, and pollution reduction comes from the Glass Containers Corporation in Dayville, Connecticut. The facility found that it could reduce particulate emmissions by increasing use of recycled glass. Under pressure from the Environmental Protection Agency to introduce pollution control measures, the plant increased its use of cullet. Today, a full 50% of its raw material comes from recycled glass—7,000 tons a year, and the firm hopes to increase that percentage. The plant now meets air quality standards without costly scrubbers, conserves fuel, produces high quality glass, reduces both litter volume and landfill requirements, and pours $2.5 million into the local economy for recycled glass.

Biological wastes are perhaps the simplest to recycle; they can be gasified, burned as fuel, anaerobically digested to produce methane, or composted and returned to the soil. In New York's South Bronx, a composting program begun in 1976 plans to restore the soil on 500 acres of vacant urban land. The Bronx Frontier Development Corporation uses vegetable wastes from a large produce market. These wastes were previously hauled to a landfill site at considerable cost. In its first year, the program restored 17 acres of vacant land; each acre can supply vegetables for 40 persons.

We have much to learn. But meanwhile, we must make choices—appropriate choices in our lifestyles, food, transportation, housing, and means of production. With a sense of purpose, we must live what we believe—by joining together with the millions of people around the world who are embracing appropriate technologies today.

Before Fritz Schumacher's death in 1977, he had seen appropriate technology programs established by the United Nations Environment Program, the World Health Organization, the U.N. Development Program, the U.S. Agency for International Development, the State of California, a dozen Third World countries, and a score of research institutes around the world. Since then the movement has continued to expand.

The author of *Present Value* has assembled an impressive testimonial to this creative impulse in one corner of the world. California has often exerted global leadership, and this innovative state is in the forefront of appropriate technology as well. We can all learn important lessons from the experiences described in this fascinating book.

Perhaps the most important lesson is the one every American school child learns while studying colonial history, but which most of us forget en route to adulthood. As children, we are incredulous that Native Americans could have sold their land and their heritage for a handful of trinkets. As gadget-crazy adults in an affluent, high-technology society, we run a grave risk of doing the same thing. We hope that *Present Value* will help open our eyes to an alternative way.

Denis Hayes
Executive Director
Solar Energy Research Institute

4

PREFACE

In recent years, California has established a reputation for innovation in a number of new technologies—including advances in electronics and engineering. Many new technologies have been developed here which offer the potential to revolutionize the way in which we use our natural resources. This publication explains and offers examples of a number of these technologies which are now in use across California. They are considered appropriate because they minimize resource consumption while providing sound economic investments.

The idea of "appropriate technology" has been advanced by a number of modern thinkers, including the late English economist E. F. Schumacher. Schumacher devoted many years of his life to encouraging what he termed "intermediate technology" in poor countries. "I have named it *intermediate technology*," he said, "to signify that it is vastly superior to the primitive technology of bygone ages but at the same time much simpler, cheaper, and freer than the super-technology of the rich." Using this perspective, Schumacher's Intermediate Technology Development Group in London developed an egg-tray producing machine for Zambia that cost $19,500; the smallest similar machine otherwise available on the international market would have cost $400,000 and would not have made the best use of native labor or resources.

Schumacher and others stimulated thinking about how mismatched technologies are to peoples' needs in the industrialized countries as well as Third World nations. Sim Van der Ryn, who formerly directed California's Office of Appropriate Technology puts it this way: "The idea of *appropriate technology* is based on a view of the world that says that industrialized societies ought to be using their remaining stock of non-renewable resources to build a society which can sustain itself when remaining stocks of minerals and fossil fuels can no longer be economically extracted and converted to use."

The foundations have been laid for that idea to build on. Since the OPEC oil embargo of 1973–74, attitudes about energy use have changed dramatically in the United States. New programs for energy conservation have been developed by government and industry alike. In 1977, California's annual electrical energy growth rate declined 50 percent from the trend established in the 1960s and early '70s. Electrical energy sales increased 3.1 percent, yet personal income increased 12.5 percent.

Even though such trends are encouraging, it is apparent that the United States must find ways to conserve energy and other natural resources through fundamental shifts in our technological perspective. American technologies have become highly productive since the Second World War, yet the energy consumption patterns in many industries have escalated at alarming rates. Agriculture is an example of a highly productive enterprise, but one which has developed an unhealthy appetite for scarce fossil fuels and other sources of energy. According to Dr. David Pimentel at Cornell University, if American-style agricultural technology were used around the world, it would offer little hope as a long-term answer to famine and food shortages. "To feed a world population of four billion while employing modern intensive agriculture would require an energy equivalent of 1.2 billion gallons of fuel per day. If petroleum were the only source of fossil energy and if all petroleum reserves were used to feed the world population using intensive agriculture, known petroleum reserves would last a mere 29 years." In many agricultural applications, up to 20 calories of fossil fuel are used to produce a single calorie of food energy.

In contrast, appropriate technologies in agriculture range from the development of substitutes for scarce fossil fuel, such as biomass fuels and solar energy, to smaller-scale farming techniques and practices which use less energy and other resources. "Integrated pest management" (IPM) is an application of appropriate technology with important implications for California's $9-billion-per-year agricultural industry. At a recent Sacramento conference on IPM technology, agricultural consultant Patrick Weddle pointed out that pear growers using IPM techniques, such as carefully monitoring orchards for diseases and minimizing chemical pesticide use, save $25 to $50 per acre per year compared to growers who use conventional, energy-intensive techniques. Weddle estimated that if pear growers in Sacramento County alone had used IPM techniques in 1976, they would have saved $300,000 in unnecessary pesticide costs and dumped 30 fewer tons of toxic pesticides into the environment. Integrated pest management is an ideal example of appropriate technology in which scientific and environmental techniques replace heavy doses of toxic chemicals. Knowledge replaces brute force—with considerable economic and environmental savings.

Another important appropriate technology, covered in considerable detail in this publication, is the use of solar energy and climatically responsive designs in homes and commercial buildings. Resource consumption in the built environment is staggering. According to the American Institute of Architects, a "high-priority national program emphasizing energy-efficient buildings" would result in saving more than 12.5 million barrels of petroleum per day, which is more than half of the current U.S. petroleum consumption. California is demonstrating its interest in reducing the excessive

6

consumption of energy in buildings through such pioneering efforts as the new state office building, Site One, now under construction in Sacramento. Site One will use a variety of passive solar design techniques to reduce resource consumption, including an enclosed court that will be daylit and will require no air conditioning; exterior window shades; a rockbed to store heat; indirect ambient lighting combined with individual task lighting; and a solar hot water system.

Many solar design techniques are not so new. In fact, research conducted by architecture professor Ralph Knowles at the University of Southern California indicates that Indians of California and the Southwest used passive solar design centuries ago. Knowles's studies of Pueblo Bonito in Chaco Canyon, New Mexico, show that the Indian dwellers of the tenth and eleventh centuries built an elegant, solar-responsive community that rivals modern designers' capabilities. The Pueblo structures are oriented toward the sun, and the building materials store heat efficiently in winter and keep the structures cool in summer. Modern climatic architectural techniques went into a period of decline following the introduction of cheap air-conditioning and heating equipment, but energy shortages and high prices are triggering a renaissance in sound design practices today.

Equally significant is the growth of the solar industry in California and other parts of the United States. In the early years of this century, a sizeable industry in southern California and southern Florida produced solar water heaters. At that time solar energy was competitive with conventional fuels—natural gas was ten times more expensive then than it is now, relative to other commodities. In the last few years, increasing energy prices have spawned a significant redevelopment of interest in solar technologies. California's state government is a leader in this national effort—the California Energy Commission has adopted a goal of assuring that 1.5 million houses and buildings are equipped with solar installations by 1985. This would result in about 20 percent of all structures in California being fitted with solar devices.

By the late 1980s, the solar industry will contribute up to $7 billion to the California economy and will add more than 50,000 jobs to the labor force. The developing solar industry is another excellent example of an appropriate technology which builds jobs in the economy while at the same time helping in a substantial way to reduce our dependence on scarce conventional fuels.

The need for appropriate technologies that consume fewer resources from the natural world but contribute to the quality of life is vital and growing. One of the major purposes of this publication is to point out that

7

adopting more logical approaches to meeting our food, energy, and water needs is economically sound for the consumer.

Equally significant is the potential impact of introducing appropriate technologies to the nation's economy. At least one leading Wall Street economist believes that a national pursuit of decentralized, appropriate technologies may be a significant path toward economic renewal. Francis Kelly, director of research at the Blyth Eastman Dillon investment firm, points out that there are two possible approaches to maintaining a high national rate of economic growth. One approach, he says, leads to the development of centralized energy and resource technologies. The other, he notes, "includes smaller scale technologies, evolutions in existing processes designed to make manufacturing pollute less and use energy more efficiently."

What's the best answer for the nation's economy, and the stock market in particular? According to Kelly, "The important thing is to restore perception of the long run. The second option could achieve that perfectly well. It would mean a recognition that the conditions for rapid economic growth—such as population growth—no longer exist. This recognition would make it possible to adapt to economic realities, rather than trying to change them as the government is now doing."

The importance of appropriate technologies is often overlooked by narrow-thinking specialists. The real lesson of appropriate technology is not confined to engineering, nor to a narrow technical discipline. By understanding the needs of today's world—to create jobs, maintain a healthy environment, and minimize the excessive consumption of natural resources—technologies can be found and developed to usher in a new era of abundance as well as permanence.

Wilson Clark

Wilson Clark
Assistant to Governor Brown for Issues and Planning

INTRODUCTION

This book presents a variety of practical, common sense methods for reducing our dependence on non-renewable resources. Such techniques have been described as "alternative" or "appropriate" technology, and rely on the use of renewable resources. They range from using solar energy to heat water to designing homes to take advantage of climatic conditions for heating and cooling. Most of these techniques are not new; solar hot water heaters were a common sight in Los Angeles homes as early as 1915. During the intervening years, however, the commercial promise of energy-conserving technologies has been dampened by the availability of cheap fuels, lack of information, and financial barriers.

The examples shown in this book demonstrate that energy saving homes, commercial buildings, and entire communities are economically feasible and perform as efficiently as traditional methods that waste dwindling resources and require a high energy and resource input.

These examples are drawn from California and, with few exceptions, have all been privately financed. Although most are tailored to a local climate, these approaches can be used in other areas with equal success. Part I shows different ways of using solar energy in homes and businesses. Part II shows how these basic concepts can be integrated and used in different structures. Whether the application is for a home or a commercial building, the entire structure is treated as an individual "energy system." Heating and cooling are provided by natural, non-fossil fuel means. Part II also looks at an entire subdivision where energy-efficient homes are built as part of a resource-conserving land use pattern. Part III describes ways an entire community can reduce its resource and energy costs with alternative approaches to food production, waste management, recycling, and energy production.

I. USING SOLAR ENERGY

Of the various energy-producing techniques that rely primarily on renewable resources, solar thermal technologies that supply heat rather than electricity have had the widest commercial acceptance. Although the state and federal government are promoting the use of solar energy, acceptance by the private sector has been slow for the following reasons: the regulated low cost of conventional fuels (oil, gas, coal, and hydroelectric power) hides their true costs, which discourages investment in solar techniques; antiquated building codes and lack of knowledge among building inspectors put some solar installations at a disadvantage, particularly when added to existing structures; building codes, banks, and lending institutions generally require conventionally fueled "back-up" units, which make solar technologies an additional "luxury" cost for the consumer. Despite such handicaps, solar technologies are cost-effective depending on how the installation is constructed, what it is used for, and what it replaces.

ACTIVE AND PASSIVE SOLAR TECHNIQUES

Active and Passive Solar Techniques. Active solar systems use mechanical devices to collect, store and distribute solar energy. Passive solar systems rely on natural convection, conduction and radiation for the transfer and storage of heat or coolness. Hybrid solar systems combine the features of both.

From the turn of the century to 1920, active solar water heaters were a common sight in Southern California. One manufacturer, the Day and Night Solar Heating Company, had sold 4,000 units by 1918. Their water heaters cut home gas consumption by 75 percent, and cost only $100 ($600 in 1977 dollars).

Now, the most widespread use of active solar devices is to heat water for domestic use and to heat swimming pools. More than 3,000 active solar units are in operation in the Los Angeles area. Solar hot water heaters range in price from about $300 for do-it-yourself units, to more than $2,000 for off-the-shelf solar hardware installed by a contractor. The State of California has a 55 percent tax credit that enables the deduction of more than half the cost of a residential solar unit and 25 percent for a solar unit on a commercial building. Lending institutions are now beginning to finance residential and commercial active solar units. The two examples described in the following sections—at Hewlett-Packard in Sunnyvale and at Hollandale Dairy in Oakdale—demonstrate how the use of conventional fuels can be substantially reduced.

Passive solar techniques use the design of a structure to take advantage of the ''thermal mass'' provided by masonry walls, concrete forms, or storage containers of water or rocks to store heat or coolness which is then used to temper a living or working space. Passive systems are most effective when an entire building is designed or rebuilt to take advantage of opportunities for passive solar heating, natural ventilation, or night cooling. Thus, it is a critical area for the involvement of lenders and builders. In the past, passive design has often implied costly custom construction, but that need not be the case. Two examples of the passive method, the Renault and Handley Solar Buildings in the Santa Clara Valley and the Savell Homes in Colton, demonstrate that this approach does not add significantly to overall construction costs and does not alter the basic appearance or function

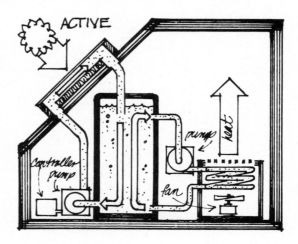

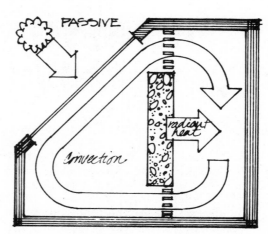

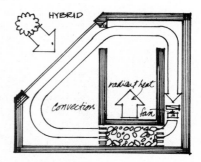

of the buildings.

Hybrid solar techniques combine elements of both active and passive systems. Usually, this involves the passive collection of energy but the active (mechanical) transport of energy using fans or pumps.

12

The Hewlett-Packard Building

Hewlett-Packard Building.
The active solar system at Hewlett-Packard's Automated Measurement Division plant in Sunnyvale, California was designed by Martin McFee, a plant facilities engineer. It is used to preheat boiler water for space heating, to re-heat cooled air, and to heat domestic water; it shows how a solar unit can improve an existing system that was not designed with energy efficiency in mind, but that cannot be replaced.

The installation consists of 10,000 sq. ft. of flat plate collectors atop a 165,000 sq. ft. building that was constructed in the mid-1960s when energy costs were low. McFee designed the unit, and six members of the maintenance crew assisted in its construction in 1973. They

built the collectors out of aluminum absorber panels, fiberglass glazing, insulation and wood frames. The panels are

connected to each other and to a storage tank with standard plastic pipe.

In this specific case, the solar unit provides heat to re-heat air that has already been cooled by a conventional air conditioning system. The solar system preheats water for a boiler that supplies hot water to 200 fan coils located in air ducts near air inlets to all rooms and work areas. A conventional chiller system sends cold air (55 degrees) through the ducts and each hot water coil adds the heat needed to reach the desired room temperature.

Hewlett-Packard's cost for the unit was about $30,000, excluding labor. From 1973 to 1976, the factory's savings on gas bills averaged $1,000 to $2,000 per month. In 1974 and 1975, savings were estimated at 65 percent of the total gas bill.

The conventional heating and cooling system in this Hewlett-Packard plant is common to many buildings constructed before the price of energy began to rise in 1973. The inexpensive solar installation that was added shows how an industry can reduce fuel bills using available personnel and expertise.

13

Hollandale Dairy

Hollandale Dairy. Hollandale Dairy is a family-operated 300-cow dairy at Oakdale, California. In this dairy, the average daily hot water consumption is 300 gallons, or one gallon per cow per day. Hot water use is divided equally between priming or warming the cow's teats before milking, and for sterilizing the stainless steel holding tank for milk. Much of this water is heated by the dairy's solar hot water system which consists of six flat plate collectors totaling 120 sq. ft. of collector area and three 120-gallon storage tanks, one of which has a gas-fueled, supplementary back-up water heater.

The predictability of a dairy's need for hot water makes this type of business a prime candidate for conversion to solar energy. The Hollandale Dairy's cows are milked on a schedule that meshes well with the solar unit. The first use of hot water is in the early afternoon, after there has been a full morning to heat water. The second use is in the evening after an afternoon of heating and storage.

Coordinating milking times with the heating cycle of the collectors maximizes the efficiency of the solar installation. Before the hot water tanks

were insulated and the solar unit was added, daily electricity consumption for cow priming and equipment sterilization was 118 Kwh. Afterward, daily consumption dropped to 69 Kwh. The installation is heating 47 percent of the dairy's hot water, with a saving of $704 per year for fuel bills.

The Hollandale Dairy solar system is a test prototype: materials and labor were donated largely by Modesto Junior College, Pacific Gas & Electric, Energy Systems, Inc., and others. Had the dairy paid full cost for materials and labor, the price tag would have been $5,050.

Increasing Energy Efficiency at Dairies

Increasing Energy Efficiency at Dairies. Regardless of how energy is provided, whether by solar or conventional means, it should be efficiently used. The following example is one way to reduce energy demands by a simple, yet effective, conservation measure.

Twenty-eight percent of the energy used by dairies is for cooling milk and heating water. Because these two processes usually take place near to each other, waste heat that is recovered from the compressor that refrigerates milk can be used to heat water with a heat exchanger. Arthur La Franchi, a dairyman in Santa Rosa, has installed a heat exchanger system that takes waste heat from two compressors and uses it to heat water in a 170-gallon tank by circulating it around the tank to warm the water inside. Water for sterilizing equipment is extracted from the top and is transferred to an electric water heater where the temperature is raised to 170 degrees; water at 80 degrees is drawn from the middle of the tank to use for washing cows.

Electricity consumption at the La Franchi dairy dropped immediately after the unit was installed. Within three months energy used for heating water had dropped 82 percent. Annual savings are about $1,337; the total cost of the unit was $1,600. Installation charges will vary, depending on the distance between the compressor and the water heater, but installed cost should be no more than $2,500.

This energy reuse technique is particularly applicable because all dairies must have refrigeration and water heating facilities, and because energy can be saved with a minimal capital investment.

How a dairy heat exchanger works

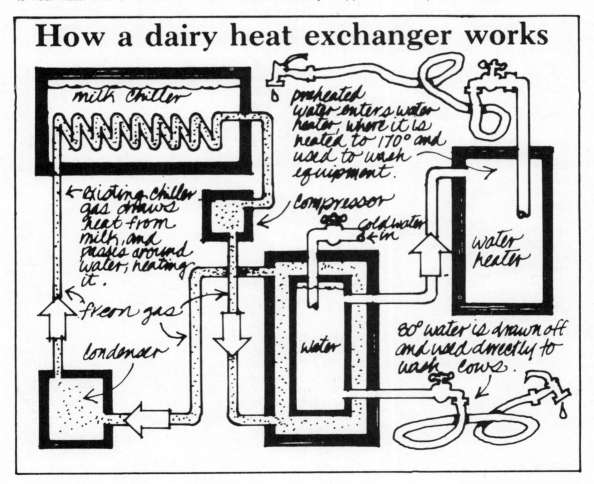

ADAPTING COMMERCIAL CONSTRUCTION FOR SOLAR

One of the most recent developments in the field of energy conserving design is the adaptation of commercial construction techniques to take advantage of natural heating and cooling cycles. The two examples here—a Trombe wall on a warehouse in the Santa Clara Valley and a cement construction system for residences —show how this has been done. The significance of these two techniques is that they use available solar energy and the structural elements of a building rather than solar collectors to use, absorb, and reuse heat.

Air inlet.

The Renault and Handley Solar Building

A commercial Trombe wall. The simplicity of the Trombe wall concept has made it an appealing passive solar alternative in new residential structures. It combines heat collection, storage, and distribution in one masonry wall, using the natural convection of hot air for ventilation and circulation. The mass required to make a Trombe wall can result in an additional construction expense in a house that does not require a thick wall for structural support. But in cases where thick concrete construction is needed, a Trombe wall system is well suited.

Harry Whitehouse, of Pacific Sun in San Mateo, designed a Trombe wall for the south-facing rear walls of two Santa Clara warehouses owned by Renault and Handley, Inc. Since the warehouses have precast "tilt-up" type panels, it was possible to adapt the structure for solar heating with small modifications in the casting forms for the panels and by adding double glazing on the outside of the building.

Casting forms were altered to include inlet and outlet openings for air flow that are spaced nine feet apart, with a vertical separation of nine feet. Concrete panels were poured and cured on site, and lifted in place with a crane. The south wall is 5½ inches thick, the same as the other perimeter walls of the building, and did

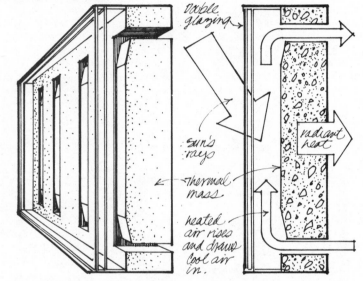

not require specialized casting procedures. The outer surface of the wall was painted flat black and glazed with a double layer of low-iron, tempered glass which is attached two inches away from the wall with standard concrete fasteners. A six-foot roof overhang shades the collector during the summer. Cool air enters the lower inlets on the inside of the building. It is forced up by natural convection and fan suction and is heated as it passes between the wall and glazing. Hot air flows through the upper openings and is distributed throughout the warehouse by a conventional fan and duct system.

The Department of Energy and the developer, Raymond

Handley, shared the cost of the Trombe wall space heating system. Because the project was recently completed, thermal and energy-saving data are not yet available. Projected performance data for the Trombe wall are shown in the appendix. Glazing and the roof overhang account for most of the cost of this design; installation cost is approximately $14 per sq. ft. of collector surface.

Normally, a substantial amount of energy is wasted in heating warehouses that are often partially open to the weather. The significance of the Trombe wall is the simple adaptability of the "tilt-up" construction technique to use available solar energy and reduce energy waste.

16

A Thermal Storage Building System

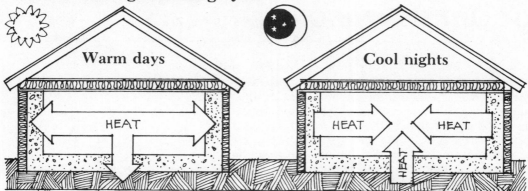

A thermal storage building system. Although concrete has been used in commercial buildings for decades, it has rarely been successfully adapted to housing construction. Jesse Savell, a developer in Colton, has overcome problems of cracking, settling and condensation with a patented construction method that results in a well insulated concrete house that costs no more than a new wood frame house. The structure can be built in 45 days, and requires only about 40 percent of the energy needed to heat and cool a wood frame house of similar size.

Savell's patented system uses 4-inch thick concrete wall panels, pre-cast in forms at a local site. A 1-inch-thick layer of polyurethane foam insulation is then sprayed over the panels that will be used for exterior walls. The insulation is covered with chicken wire, and then a thick layer of stucco is applied. A concrete slab house foundation is poured and insulated around its perimeter. The interior and exterior wall panels are then tilted-up and set on the slab with the insulation facing out. Walls are not rigidly joined to each other, to the roof, or to the foundation. Thus, each panel is free to move slightly without creating stress in the overall structure that could cause cracking.

The net result is a conventional-looking house that has an average thermal mass of 90,000 pounds in direct contact with the living space but is insulated from the outside environment. Contact between the earth and slab allows the earth to act as a heat sink, helping moderate temperatures within the house. On hot days, excess heat is absorbed into the massive concrete walls, maintaining a cool temperature without air conditioning. At night, the heat the walls have retained is radiated back into the living space.

In winter, the house is protected by its cocoon of insulation that can be made as thick as the climate requires. For additional winter heating, the interior walls are warmed by the sun on a clear day or by excess heat from lights, appliances, space heaters, or the occupants. Owners of Savell homes near Los Angeles report that when outdoor summer temperatures are 95 degrees or higher, interior temperatures are about 76 degrees. During mild winters, the interior temperature does not drop below 72 degrees. A stabilized environment of this kind requires virtually no supplementary heating or cooling.

This approach to home building is flexible enough to meet the needs of custom designers as well as mass-production builders, and generally results in construction costs that are comparable to those of standard wood frame dwellings. In addition it satisfies all building codes and standards and meets conventional financing requirements. The Savell method has not yet been proven in extreme climates, but homes being built in Wyoming should indicate whether that type of structure is suitable for a cold climate. Structures capable of thermal storage (storing heat) may be a key to low-cost solar homes.

SOLARIUMS AND GREENHOUSES

Solar heat that collects in a sun room (solarium) or greenhouse can also be used to warm a house if such structures are attached to or integrated with the house. Heat must be stored in a thermal mass—usually in rocks, cement, water or the mass of the house itself—and air must be circulated from the solarium throughout the house either mechanically or by natural convection. Induced air circulation will ventilate and cool a solarium. Use of a high vent and low vent will create a "stack effect," as heated air is pulled up and out the top vent.

The benefits of a solarium are not limited to heat collection. Sun rooms have been built on houses for decades, providing natural light and warmth regardless of how the room is used. Designs vary in complexity from a simple, attached greenhouse to complex shelters for heating, cooling, raising fish and growing food. The following examples—a house with a solarium integrated into the original design and an existing house that had a solar greenhouse added—indicate the potential of such a structure for providing heat, light, and food.

Integrating a Solarium into a New Home

photo by Susan Jean Benson

Integrating a Solarium into a New Home. Peter Calthorpe, a San Francisco designer, has designed a house that uses a solarium to produce food, provide heat, serve as a vestibule and a circulation area, and has a stairway inside that joins the two floors of the house. This house is a good example of a hybrid solar system that makes use of various mechanical devices to circulate passively collected heat.

All major rooms of this U-shaped house open onto the solarium, thereby receiving natural light and heat. The main entryway through the solarium acts as a buffer, reducing infiltration of cold air into the living space. A stairway rises through the greenhouse to the rooms on the upper floor, eliminating the need for costly and space-consuming stairways inside the house.

The thermal mass in this house is located under the slab floor of the living room. Air warmed in the greenhouse rises to the peak and passes through what Calthorpe calls a "supercharger," a 65 sq. ft. flat-plate air collector. Here the air is heated to about 100 degrees and enters a plenum, where a fan forces it down to the rock bed under the living room floor. Warmth radiates into the living room through the slab floor and through floor registers. Cool air sinks to the bottom of the rock bed where a fan forces it back into the greenhouse to be reheated.

For summer cooling, the glass cover over the supercharger is opened. Air rising up is superheated as it passes over the black absorber plate and is forced out, drawing the air below it upward. Low vents are opened to provide a continual supply of air and to cre-

ate a stack effect. To augment air flow throughout the house, windows on its north side can be opened, providing cross ventilation.

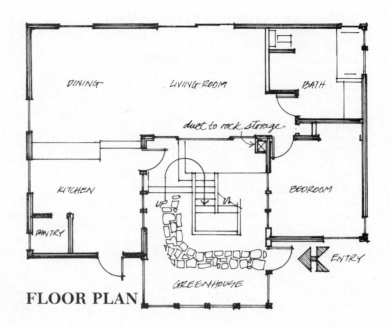

FLOOR PLAN

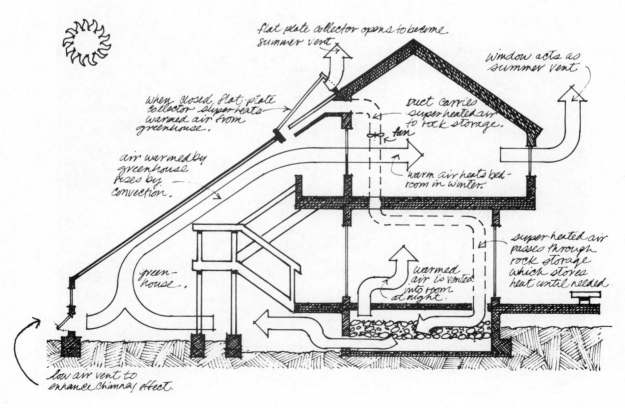

19

Adding an Attached Greenhouse

Adding an Attached Solar Greenhouse The addition of a greenhouse to the south-facing side of an existing house is termed a "passive solar retrofit." Basic passive techniques already described are used; but, in this design, an enclosed, glazed greenhouse is the primary collector of solar energy. Homes with unobstructed southern exposures are often well suited for a greenhouse retrofit. In this example, a greenhouse was joined to an older, split-level, Spanish-style house in Sacra-

mento. Design and construction of the greenhouse was a joint project of the homeowner Jeff Reiss and Lynn Nelson, director of the Habitat Center of San Francisco. The Habitat Center is a non-profit educational institute that sponsors weekend greenhouse construction workshops and promotes the passive use of solar energy.

Reiss' greenhouse is a particularly good example of the potential of attached solar greenhouses. Basic conservation measures such as an in-

sulated ceiling and exterior shading reduced the heating and cooling problems. In addition, the compact, split-level design of the existing house provides efficient air circulation throughout the house. The greenhouse was designed to complement the existing architecture.

The 21 ft. by 7½ ft. greenhouse is designed to provide winter heating and summer cooling. In the winter, 12 black, water-filled, 55-gallon drums in an alcove absorb heat. Heat is released into the house on all three floors through a door, a window, and high vents cut

into the wall to let warm air into the upstairs bedroom. Circulation takes place by natural convection. When heating is unnecessary, all the openings into the house are closed off and vents in the greenhouse are opened to release hot air.

In the summer, temperatures in the greenhouse are reduced by the overhang, which completely shades the south wall and the barrels. The angle of the double pane glass is steep to reflect the high sun and reduce heat gain inside the greenhouse. The low vent on

the west and the roof vent are opened, creating air circulation

as warm air is drawn up and out of the greenhouse. Thermal chimneys exhaust hot air from the upstairs rooms and draw cool air from beneath the house.

The cost of materials for the 200 sq. ft. greenhouse was approximately $1,600 or $8 per sq. ft., excluding the cost of labor and permits. Reiss estimates that the greenhouse will provide more than 60 percent of the heating needs of the house, and the thermal chimneys will provide all of the cooling needed in summer.

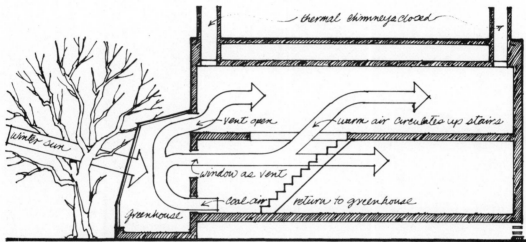

Winter heating

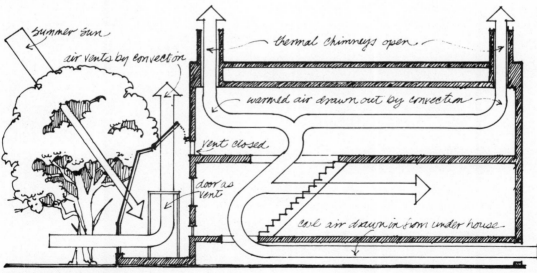

Summer cooling

II. SOLAR ENERGY FOR MANY CLIMATES

The following case studies were made of houses and commercial buildings in which some of the approaches already described are integrated. Each illustrates methods of designing for California's wide range of climatic conditions. The architects designed these structures to complement the local climate, to conserve energy, and to reduce the need for supplementary heating and cooling.

As of July 1, 1978, all new buildings are required to meet energy conservation standards established by Title 24 of the California Administrative Code. Title 24 establishes both prescriptive and performance standards. Buildings that cannot meet prescriptive standards, which specify how a building must be constructed, must demonstrate that they perform equally as well. Performance standards establish criteria limiting the annual energy consumption or the hourly rate of a building's heat loss (measured in Btu's per sq. ft. per hour).

Although most of the buildings examined in this chapter were constructed before Title 24 went into effect, they are typical of the kind of structures that would have to meet current thermal performance criteria. In each case study, only the total cost and the cost per square foot are given. Climatic variables, construction details and financing information are presented in the appendix.

The Greenstein residence

In the San Fernando Valley the challenge is to design a comfortable, energy-conserving home that takes advantage of the warm days and cool nights and provides adequate air circulation and ventilation. Rob Quigley of Gluth and Quigley Associates, San Diego, designed a home for the Greenstein family that combines separate passive heating and cooling techniques in a house which is actually a thermal chimney.

The Greenstein residence is in Woodland Hills, northwest of Los Angeles. In winter, daytime heat is seldom required, but the evenings are another matter. To take advantage of this temperate climate, the architect integrated four primary energy techniques in the Greenstein residence. Of these, the most unique is the ventilation system.

Quigley designed the 1,600 sq. ft. house around a 10-foot-square and 25-foot-high belvedere. With the exception of the master bedroom and the children's room, all the rooms open onto the belvedere. Two large industrial turbine ventilators draw air through the house and exhaust it through the peak of the belvedere, which creates a chimney effect—a constant suction of air through and out of the house. At the

base of the turbine ventilators are motorized louvers that can be opened or closed. If additional cooling or heating is needed, the chimney exit is blocked off by a large heavy canvas panel, like a damper, that swings down and covers the top of the belvedere. Thus, the house is designed to

reduce the space that has to be heated or cooled.

For heating, the furnace fan draws household air through a south-facing, glazed 8 ft. by 12 ft. alcove past black, water-filled oil drums. The heat that has been absorbed into the drums is released to the air and distributed through a standard system of ducts and floor registers. Quigley has designed the installation so that it uses the fan of the forced-air gas furnace, required by building codes and by the lender as a condition for financing. A thermostat monitors the air temperature, and if the air is not warm enough after it passes through the waterwall, the furnace turns on. At night, an insulated door is drawn over the waterwall to prevent heat loss.

When summer temperatures

24

soar and the thermal chimney cannot adequately ventilate the house, the Greensteins can resort to an additional cooling system. Here is how it works: at night, the furnace fan draws cool outside air over a bed of rocks, cooling them, and discharges the air back to the outside. During the day when additional cooling is needed, the outside openings are closed off and the furnace fan draws household air through the cool rock bed and recirculates it through the house, using the same duct system as for heating.

The architect has also installed a standard solar water heater to heat water for domestic use. In addition to these major components, he has used other techniques, such as exterior shading on the south and west sides to reduce the amount of heat the house absorbs.

Quigley believes that an energy-efficient house need not cost more than a conventional residence. The techniques he used are integrated into the Greenstein residence so that they are aesthetically pleasing and economical.

The Woodland Hills residence meets all applicable building codes. Total cost was $73,600, or $46 per sq. ft., which is comparable to most new construction in the Los Angeles area.

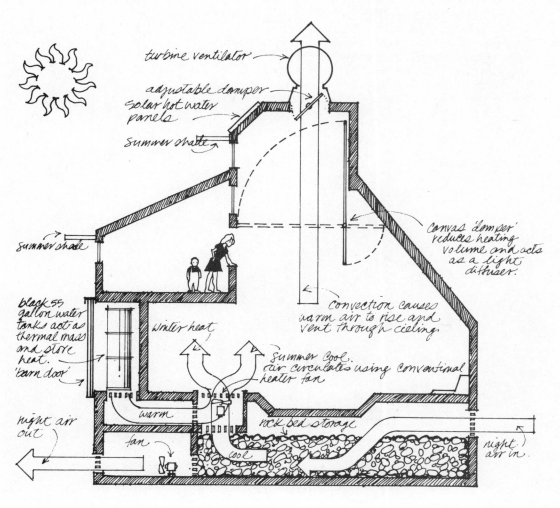

turbine ventilator

adjustable damper
solar hot water panels

summer shade

summer shade

canvas 'damper' reduces heating volume and acts as a light diffuser.

black 55 gallon water tanks act as thermal mass and store heat.

'barn door'

winter heat

convection causes warm air to rise and vent through ceiling.

Summer cool air circulates using conventional heater fan

night air out

warm

fan

cool

rock bed storage

night air in.

25

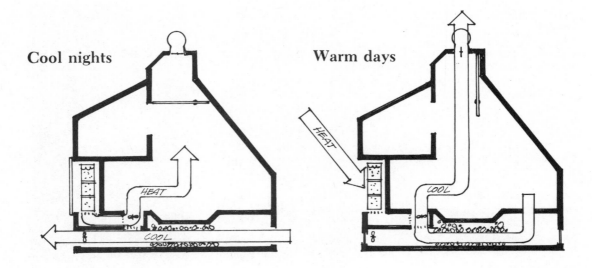

Cool nights

HEAT

COOL

Warm days

HEAT

COOL

The Greenstein residence represents the type of architecture that is appropriate to the climate of California's most highly populated region. Quigley's primary energy-conserving device—the thermal chimney—is an integral part of the structure. The conventional energy back-up system—furnace and duct work—is used for heating and cooling, taking advantage of units the owner had to install to secure financing. The entire house was built with standard methods which eliminate unnecessary costs and require no specialized construction techniques.

The Tom Smith residence

Tom Smith had eight criteria in mind when he asked Lee Porter Butler, a San Francisco architect, to design an energy-conserving home for the Lake Tahoe area using a passive solar system. His criteria were to: (1) achieve 80 percent self-sufficiency for heating and cooling; (2) conform with existing building codes; (3) meet standard financing requirements; (4) match the construction costs of a standard home; (5) avoid mechanical systems or unusual technologies that spoil the appearance of the house; (6) satisfy conventional taste in architecture and design; (7) provide a healthy and pleasing environment for the occupants; and (8) employ standard technology and standard materials in construction. Together, Butler and Smith designed and built a house that meets all of these requirements.

Smith chose Lake Tahoe as the site for this 1,800 sq. ft. home. The cold Sierra Nevada winters would test whether the criteria could be met using a greenhouse or solarium as the primary heat collector, natural convection to circulate heat, and a design which enables a blanket of warm air to isolate and insulate the living space from outside climatic variations.

The design concept was simple: If the living space is surrounded by a protective envelope that buffers it from the fluctuations of the outside temperature, then the heat released by lights, residents, and radiation from the inner walls should be sufficient to maintain a comfortable interior temperature, even on cold days.

The energy system that provides the protective envelope starts at the greenhouse, where air is heated. Warm air rises by convection into a 12-inch space between the interior ceiling and the roof of the house. The air flows down between the back walls, warming the inner shell of the living area, and then moves under the house, where a thermal mass of loose backfill absorbs any remaining heat. The air flows toward the greenhouse again, drawn up through ¼-

27

inch spaces in the greenhouse decking by the rising hot air above.

At night air circulation slows, forming a warm air blanket around the house, and as the temperature in the greenhouse drops below that in the rest of the envelope, the circulating air reverses its cycle. During cold weather the greenhouse can be blocked off from the living areas allowing its buffering and heating function to continue while isolating the living areas from external variations in temperature.

The greenhouse has a floor area of 300 sq. ft. and is wrapped in 390 sq. ft. of double-pane glazing. Smith reports that a house of this design need not face south to be energy efficient, pointing out that his residence is 20 degrees off true south.

Heat is stored in two places: in the well insulated internal mass of the house itself and in the backfill beneath the house. Both sources store enough heat to keep the house comfortable for up to two weeks, even with generally cloudy conditions. During the winter, the greenhouse temperature never went below 52 degrees. When necessary a wood-burning stove warms the living room.

At Lake Tahoe's 7,000-foot elevation, cooling is rarely a problem. A row of clerestory

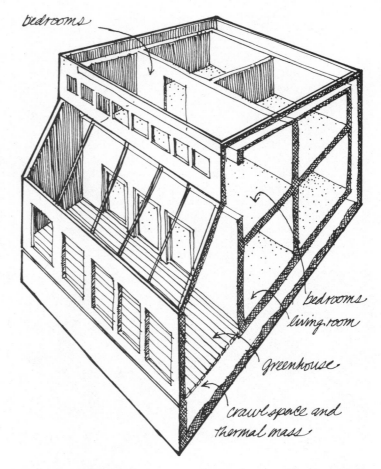

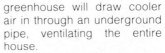

bedrooms
bedrooms
living room
greenhouse
crawl space and thermal mass

windows at the top of the greenhouse can be opened to let hot air escape. An underground pipe leads to the crawl space on the north side. Hot air escaping from the open clerestory windows at the top of the

greenhouse will draw cooler air in through an underground pipe, ventilating the entire house.

To qualify for a loan and to meet the local county building codes, Smith installed a small electric baseboard heater. It has never been turned on. In February, the coldest month of the year, the electricity bill for appliances and an electric hot water heater was $38, which was $142 less than the neighbor's bill for a house of similar size.

The branch of Bank of America in Truckee financed construction of the house. It was built for $53,258 which included the contractor's profit. The overall cost per square foot was about $30, considerably less than the average Lake Tahoe construction cost of $37 per sq. ft.

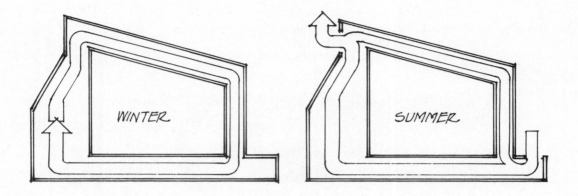

WINTER   SUMMER

The basic design of Smith's house can be modified for warmer climates: greenhouse vents, overhangs, attic fans and shaded window area can help prevent overheating. In addition, the same passive approach can be used in homes with a different design. The unique qualities of this house are the elimination of costly mechanical systems, its efficient use of energy, and a design that meets conventional financing requirements that can be built by local builders.

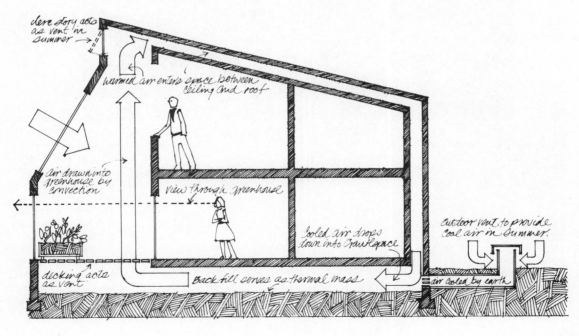

29

The Dave Smith residence

Until recently, single family homes designed to suit a specific climate were relatively expensive. This is no longer true, and architects and builders are producing a wide variety of energy-conserving homes in California. Dave Smith, a Sacramento contractor, built his own duplex that combines elements of both active and passive systems without relying on unusual designs or unconventional construction methods.

The duplex was designed by Sacramento-area architect Brent Smith, who is known for homes that make efficient use of energy and space. The duplex has approximately 2,500 sq. ft. The small unit, 1,100 sq. ft., has one bedroom, a guest room, and a large living space; the adjoining larger unit has two full-sized bedrooms. Other than minimal interior changes, both units have the same basic plan and include the same en-ergy-conserving features—a solarium, and solar space and water heating.

The solariums and garages face south. Two rows of commercially available flat-plate solar collectors on the garage roof heat water, which continually circulates through the 256 feet of collectors, into the 1,066-gallon tank and back to the collectors again in a closed system. Water for domestic use flows through a coil in the main tank where it collects heat and returns to a standard 40-gallon, gas-fired water heater for storage and use. If the solar unit does not get the water hot enough, the gas turns on to produce the required temperature.

Each unit has an 80 sq. ft. solarium which collects warm air to help heat the rest of the house. When the double-glazed French doors at the entrance to the solarium are

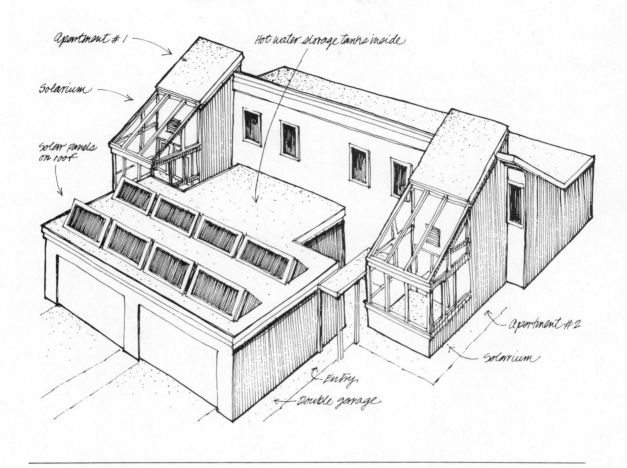

Apartment #1

Hot water storage tanks inside

Solarium

Solar panels on roof

Apartment #2

Solarium

Entry

Double garage

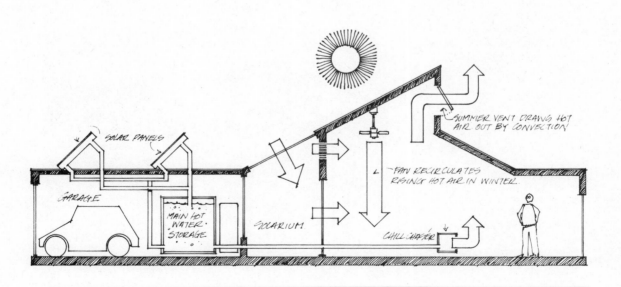

SOLAR PANELS

GARAGE

MAIN HOT WATER STORAGE

SOLARIUM

CHILL CHASER

FAN RECIRCULATES RISING HOT AIR IN WINTER.

SUMMER VENT DRAWS HOT AIR OUT BY CONVECTION

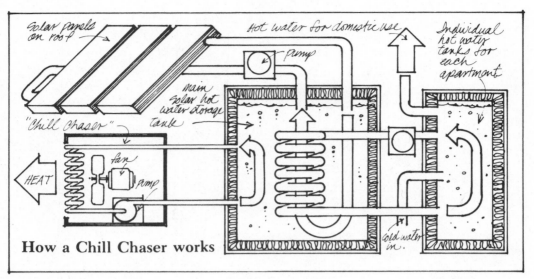

How a Chill Chaser works

Labels in diagram: Solar panels on roof; Hot water for domestic use; Individual hot water tanks for each apartment; pump; main solar hot water storage tank; "chill chaser"; HEAT; fan; pump; cold water in.

opened, warm air circulates freely. A small fan at the peak of the solarium pushes air out and into the living and dining area.

A slowly revolving ceiling fan provides additional air circulation and keeps warm air from collecting in the clerestory ceiling. The exposed slab floor of smoothed, polished aggregate throughout the house and solarium stores heat.

The solar space heating method in the Smith's duplex is relatively simple. Hot water is pumped from the top of the main tank beneath the slab foundation to a "chill chaser" —a small unit in the wall that circulates hot water through a coil in a radiator. A fan blows air across the heated coil and into the room in cold weather. A wood-burning stove provides additional heat.

Insulating shutters protect the north wall and insulating curtains keeps cold air from filtering through the double-pane windows throughout the house. Smith has framed the sliding glass door on the north side in a valance to keep the cold from circulating around the perimeter of the curtain and into the living room. Shutters and curtains also help keep heat out in the summer.

To reduce the temperature in the solarium, Smith hung exterior shading over the roof and closed the double-glazed doors across the entryway. Vents serve to circulate air out of the solarium. The west side is shaded by plants and exterior bamboo shades. There are no conventional space-heating back-up systems in this duplex.

Dave Smith reports that a home of this type could be built for $80,000, or $32 per sq. ft. The actual cost of his duplex was less because he built it himself.

A chill chaser.

The significant feature of this house is that it is a multiple family dwelling in which the energy devices are unobtrusive and easily maintained. Such installations would fit into any neighborhood and be acceptable to any resident.

The Harmon residence

In towns like Barstow, El Centro and Bakersfield, California, endless rows of houses are cooled, at great expense, by electric air conditioners. Although summer temperatures reach 125 degrees, passive solar heating and natural ventilation are still feasible. Jim Harmon, a political science professor at San Diego State University, and Peter Hansen designed and built this desert house using simple, time-proven passive cooling techniques similar to those used by the southwest Indians. The house uses minimal energy, is comfortable year-round, and harmonizes with the desert environment. Harmon's house is the only one in this book that required special construction techniques, mainly because of the temperature extremes and its unique earth-sheltered design.

Harmon's house is 25 miles west of Calexico, near the California-Mexico border. The region has a typical desert climate, with normal humidity of 10 percent, summer highs of 125° F, and winter lows of 30° F. Harmon uses a variety of techniques to keep his house comfortable throughout the year without conventional heating or cooling systems. The octagonal house is built into a low hill; on the south, west, and north sides, it is buried five feet underground. The east side provides a view of the surrounding desert, and eye-level windows on the other sides provide light and scenic

views. Prevailing winds from the west blow over the side of house with the berm protecting the entrance on the east side.

All the features of this house were designed for their cooling effect. The 1,000 sq. ft. foundation is 5 feet below the surface, which stabilizes the temperature inside the house. The walls are of waterproofed concrete up to ground level. All exposed walls are built like a thermos bottle. From the inside working out, the layers are: interior walls; two layers of fiberglass batt insulation; plywood; one-inch-thick, foil-backed styrofoam insulation; an air space;

and exterior walls of Mexican bricks.

Structurally, the interior ceiling resembles a horizontal eight-spoked wheel. The spokes and center axle are made from railroad bridge timbers. Where the timbers meet in the center, a series of vents lead through a tower to the outside.

To ensure adequate ventilation, the Harmon design enables heated air to rise up through a double roof and out the tower, drawing cooler household air out with it.

At the edge of the tar and gravel/plywood roof are small holes. Air enters here and is heated as it passes under the

33

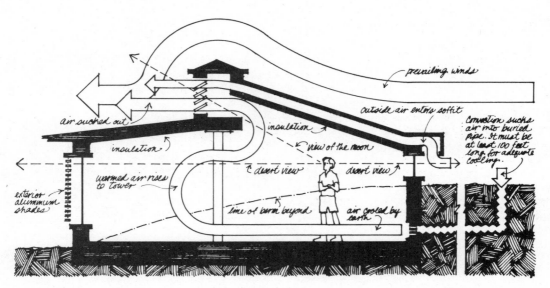

Labels in diagram:
prevailing winds
air sucked out
insulation
insulation
view of the moon
outside air enters soffit
Convection sucks air into buried pipe. It must be at least 100 feet long for adequate cooling.
exterior aluminum shades
warmed air rises to tower
desert view
desert view
line of berm beyond
air cooled by earth

roof. Air inside the house is entrained by the flow of hot air above it, and is drawn out the vents in the tower. Below the air space are two rafters, separated by multiple layers of insulation.

A passive air conditioning system supplies cool air to the house through eight corrugated tubes buried deep in the ground. There is a single terminal 100-feet from the house, and the tubes connected to it lead to floor registers on each of the eight sides. Air flows in at the terminal and down the tubes, where it is cooled by the 70 degree temperature of the

earth. The air is pulled upward into the house by natural convection created by hot air moving across the roof and out the tower: Hot air leaving the tower sucks the cool air out of the earth tubes, into the room and then out the tower. If additional cooling is necessary, Harmon has installed a swamp cooler over the air terminal and a 30-inch exhaust fan in the tower.

The walls are so thick that heating is not a problem, even in the coldest weather. Long windows on the southeast side are protected by an overhang, which keeps sunlight

out of the house from March to October. Harmon has installed exterior blinds to keep direct sunlight from hitting the

Hot air is exhausted out this tower.

34

windows at any time of year. Skylights on either side of the tower provide additional lighting and a view of the moon as it crosses the desert sky at night. An overhang protects these skylights from direct sunlight.

To increase humidity inside the house, Harmon added a Japanese bath. Hot water for the tub is heated with solar collectors. After being used on winter evenings, the bath is not drained, which adds to the humidity. In the morning, the water is drained and diverted to the garden.

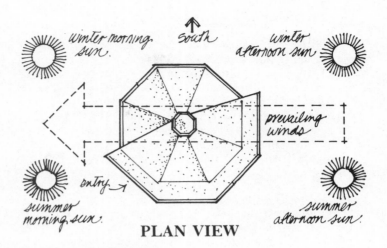

PLAN VIEW

Although this particular design and construction is applicable to only a few areas in California, the cooling and ventilation methods can be used in other climate zones. Harmon has naturally heated and cooled his house with minimal disruption of the environment. From the road, only the peak of the tower is visible. Native vegetation requiring little maintenance and water has been planted around the house. The unobstrusive coloring of the house blends into the desert landscape. Total cost of construction was $22,000 or $22 per sq. ft., including labor.

The Hodam residence

Natural techniques for conserving energy and heating and heating and cooling buildings characterize most of the examples in this book. Of equal importance, however, is the integration of food production, waste recycling, and water conservation. This integrated approach was used by Bob Hodam when he designed and built his 2,800 sq. ft. house near Placerville. His goal was to combine a variety of techniques for conserving and reusing resources as a natural part of the conservation-oriented lifestyle of the residents.

Site for the house is the Sierra Nevada foothills. Winters are generally mild with an occasional snowfall. The three-story, wedge-shaped house makes use of daily temperature fluctuations for cooling and ventilation in the summer and has a greenhouse for heating in the winter.

The need for supplementary heating and cooling was reduced by thoroughly insulating the house and by orienting it in the direction of the prevailing breezes. The greenhouse and 70 percent of the total window area are on the south side. When these windows are open southerly breezes circulate through the house on all floors.

In winter, hot air that collects in the greenhouse is distributed through the house primarily by natural convection. Large sliding glass doors separate the greenhouse from the lower floor. When they are opened, warm air enters the house and rises to the upper floors through the stairwell. A row of plexiglass hatches separates the second and third floors. When they are closed, warm air remains in the second floor; when they are opened, the warmth filters upward into the attic work space. A doorway into the greenhouse from the landing of the stairwell can also be opened to admit warm air.

Supplementary heating is provided by a reverse cycle heat pump. Normally a five-ton heat pump would be required to heat or cool a house of this size, but because conservation and other techniques are used to satisfy most of the heating needs, a 2½-ton heat pump was installed.

Summer cooling is accomplished by convection and natural ventilation. Doors leading to the greenhouse are closed and curtains are drawn across the downstairs windows to block out heat. Windows are opened on the north side of the lower floor to assist air circulation. Cool air enters, rises through the upper floors, and exhausts out the north windows on the top floor. As evening temperatures begin to drop, open windows catch the breezes.

Hot air is most likely to collect in the top floor. Small wing windows on the east and west sides act as scoops, drawing in the evening breeze. In the greenhouse, Hodam is testing different methods of regulating the amount of heat the greenhouse picks up during the day. The west half has been glazed with double-pane glass. The east half has been covered with insulated shutters. On par-

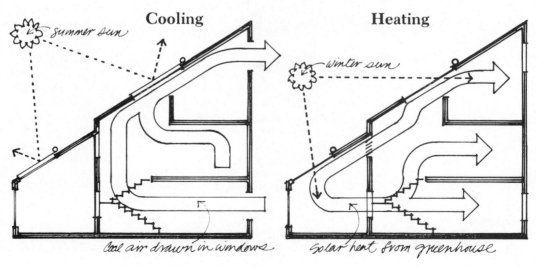

Cooling — *summer sun* — *Cool air drawn in windows*

Heating — *winter sun* — *Solar heat from greenhouse*

36

ticularly hot days, the reverse cycle of the heat pump is used for air conditioning.

Hodam is experimenting with a combination of hydroponics and aquaculture in the greenhouse. Hydroponics is the culture of plants in a non-soil medium, usually washed sand, that is periodically irrigated with a nutrient-rich solution. Around the perimeter of the greenhouse is a two-tiered row of trays. In the upper tray are the hydroponic beds. The lower tray will be used to grow catfish or freshwater shrimp.

When additional heat is needed, hot air from the nearby clothes dryer can be recirculated through a duct into the greenhouse.

Throughout the house, there are flow restrictors, low-flush toilets, and water conserving showerheads. The Microphor toilet uses two quarts of water per flush and relies on compressed air to force waste into

the discharge line. Air is pressurized by a 1,150-watt compressor that operates about one minute a day (using about 10 cents' worth of electricity a year). The Minuse showerhead has a 400-watt blower to create a fine spray adequate for showering, and uses about one-sixth of the water of a standard showerhead.

These water conservation devices save energy and money. Water is conserved, and the cost of heating it is about $18 per year, or one-fifth the annual cost of a conventional system.

Another feature enables

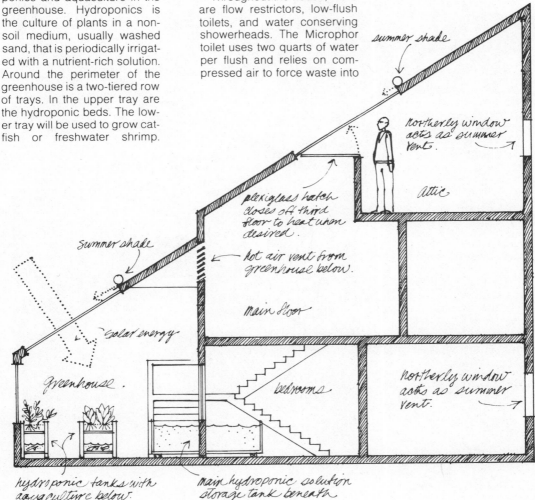

summer shade

northerly window acts as summer vent.

attic

plexiglass hatch closes off third floor to heat when desired.

hot air vent from greenhouse below.

main floor

summer shade

solar energy

greenhouse.

bedrooms

northerly window acts as summer vent.

hydroponic tanks with aquaculture below.

main hydroponic solution storage tank beneath stairway.

AN INTEGRATED SYSTEM

37

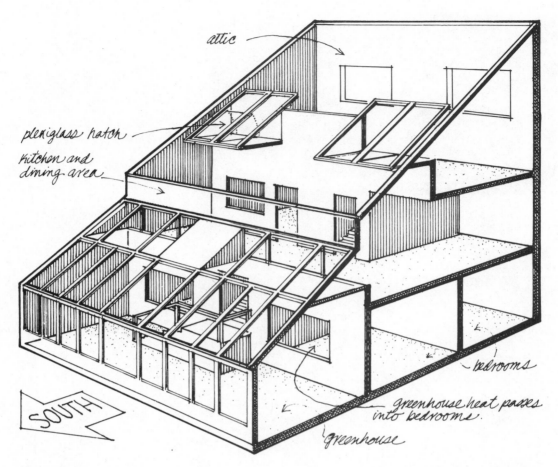

attic

plexiglass hatch

kitchen and dining area

SOUTH

greenhouse

greenhouse heat passes into bedrooms.

bedrooms

some wastewater to be reused in the garden. The water discharge lines from the kitchen sink, bathroom shower, sink and laundry have been kept separate from those for the toilet until they approach the septic tank. The sink and laundry line has a surface outlet that is currently capped, but is potentially a source of greywater for irrigation .

The total cost of this house was $90,000, including the greenhouse. Primarily because of the unique features in the house, building costs were higher than those generally considered acceptable by local banks. Despite this, construction costs, including labor, were $33 per sq. ft.

38

The Terman Engineering Center

Palo Alto has a Mediterranean climate, with warm days, cool nights, and gentle breezes that inspired the designers of the original Stanford University Quadrangle. The architectural style is characterized by imposing arches, long shaded arcades, stucco masonry, and overhanging red tile roofs. These structures were built to complement their environment and climate. Since the original campus plan was conceived, this climate-responsive architectural style was abandoned. With the completion of the Frederick Emmons Terman Engineering Center in 1976, the tradition has been reintroduced in a way which both complements the original Spanish architecture and saves energy.

photo by Richard Dawson

The dominant features of the Terman Center are the natural ventilation and lighting system and the wood and stucco construction. Various factors influenced the decision to rely on a natural, rather than a mechanical, cooling and ventilation sysem. The Dean of the Engineering School was interested in a sound, energy-conserving style of architecture. Also, the building was planned in early 1974 when energy awareness was at a high level, steel and concrete were expensive, and active solar technologies had not been tested on large buildings. In addition, students, staff and faculty had jointly agreed on the characteristics they wanted the building to have.

Among those deemed important were adequate air movement and acceptable air quality even when rooms are full; minimal need for energy-consuming lighting; natural light; windows and skylights shielded from sun; and offices with windows on the outside of the building. These requirements resulted in the selection of Harry Weese, a Chicago architect with extensive experience using natural energy systems, to design the center.

The outcome was a structure that met most of the criteria of the engineering school and of the University, while not exceeding the $9.2 million budget.

The Terman Engineering Center is a seven-story, 150,000 sq. ft., L-shaped building with a pool on the south and a shaded sunken garden on the north. The first two floors are made of pre-cast concrete. The floors above the concrete base are framed in wood; laminated beams of fir and hemlock are bolted to fir columns. The area between the columns is finished with brushed stucco. Because the structure is naturally ventilated, no duct work is required. The interior ceilings expose the structure of the floors above. All of the building's service systems— sprinklers, water pipes, cable trays—are exposed, a feature that saves money and was considered desirable by the users.

All offices have floor-to-ceiling French windows with bronze-tinted glass. Outside the windows are hand-oper-

39

ated louvered shutters that slide closed on tracks. When closed, the shutters block out direct sunlight but allow soft filtered light and air flow through the rooms. The corridor doors are made with an insert which, when opened, allows air movement through a louvered door panel. This feature enables visual privacy while allowing for air circulation. Each room is equipped with baseboard hot water convectors for winter heating.

Vertical shafts in the corridors are topped with operable skylights. These induce circulation by exhausting hot air and simultaneously reflecting light down the shaft into the corridors, minimizing the need for corridor lighting.

The combination of natural lighting and ventilation methods has proven effective. Two summers have passed, and even those who strongly advocated air-conditioning during the planning stages admit that the Terman Center is a pleasant place to work. One professor noted that the induced ventilation, due to the stack effect created by the open skylights, is so effective that he has trouble keeping messages tacked to his bulletin board. He also reported that, after working in the building two years, he has yet to turn on the heat.

Portions of the building have conventional air-conditioning —laboratories, the library,

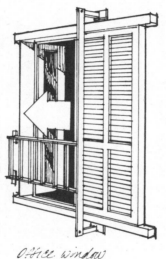

office window

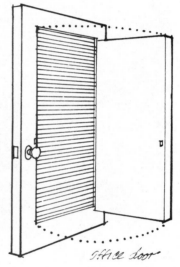

office door

photo by Richard Dawson

some classrooms, and the auditorium. When the building was opened, the only problems were with the air-conditioning equipment. The Terman Center is less costly to operate than other buildings of comparable size and use on the Stanford campus. Details related to energy savings are shown in the appendix.

The Terman Engineering

Center was completed in 1976 for $55 per sq. ft., well under the projected cost of $59 per sq. ft. Savings realized by using wood for the major structural elements were spent to enhance the fine details of the building; for example, holes where beams are bolted together have been filled with wood plugs and the louvered doors are custom-milled.

40

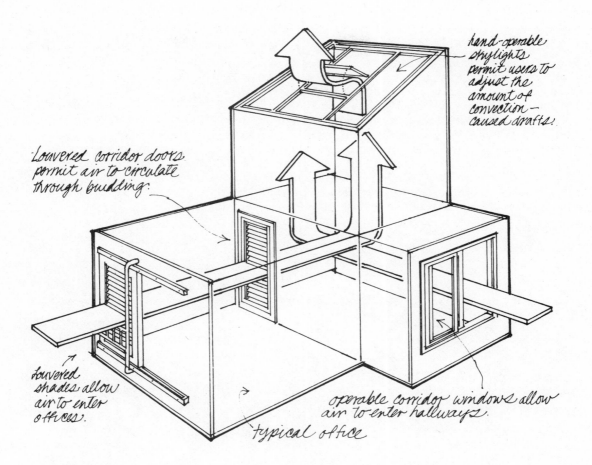

hand-operable skylights permit users to adjust the amount of convection-caused drafts.

Louvered corridor doors permit air to circulate through building.

Louvered shades allow air to enter offices.

operable corridor windows allow air to enter hallways.

typical office

Natural ventilation is feasible in many parts of the state, particularly Southern California. The pleasing qualities of the Terman Center as a work place and its economical use of energy should resolve any doubts about the feasibility of a naturally cooled structure designed to fit specific climatic conditions.

photo by Richard Dawson

41

Site One

A new state office building is under construction in downtown Sacramento that is designed specifically for the climate in which it is being built. The four-story office structure is called Site One because it is the first of several buildings being constructed under the state's Capitol Area Plan. The goal of the designers is to reduce energy consumption—Site One is expected to use 40 percent less energy than a conventional structure built to meet California's Title 24 non-residential building conservation standards—and to provide a comfortable and pleasant work environment for state employees.

A striking feature of the 276,000 sq. ft. building is a central atrium or court. The landscaped court replaces corridors as the main passageways and includes areas where employees can meet for lunch and coffee breaks. While most buildings have hidden, inaccessible stairways that are hard to find and use, Site One will have two highly visible stairways leading from the court to the floors above. In addition to improving the work environment with plants, light, and space, the court has a number of energy-related benefits. North- and south-facing skylights cover the court,

providing light and air. Exhaust fans and louvers that shade the skylight will be computer-controlled to maintain a comfortable environment while using only a fraction of the energy that would ordinarily be required to heat or cool that much space.

During winter, the exterior shading louvers can be opened to allow direct sunlight to enter the court, thereby providing ample passive solar heating. In summer, the louvers will automatically block the sun's rays, while admitting indirect daylight. During the day, the court will remain cooler than the outside because the

direct sun is blocked out. Most of the infiltrated heat will be absorbed into the concrete mass of the structure. On summer nights, the building will take advantage of Sacramento's cool breezes from the Delta. Six exhaust fans at roof level will flush out the heat that builds up in the court's structure, and pre-cool it before the next day.

Some of the same principles are used to reduce energy demand in other parts of the building. Most of the building's concrete structure will be exposed, allowing it to absorb part of the radiant and convected heat given off during the day by lights, employees and

42

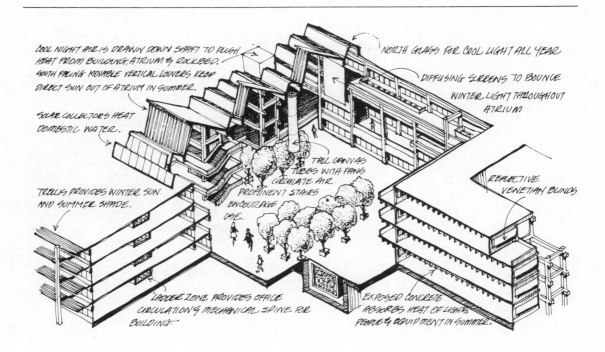

COOL NIGHT AIR IS DRAWN DOWN SHAFT TO FLUSH HEAT FROM BUILDING, ATRIUM & ROCKBED. SOUTH FACING MOVABLE VERTICAL LOUVERS KEEP DIRECT SUN OUT OF ATRIUM IN SUMMER.

SOLAR COLLECTORS HEAT DOMESTIC WATER.

TRELLIS PROVIDES WINTER SUN AND SUMMER SHADE.

TALL CANVAS TUBES WITH FANS CIRCULATE AIR. PROMINENT STAIRS ENCOURAGE USE.

'LADDER' ZONE PROVIDES OFFICE CIRCULATION & MECHANICAL SPINE FOR BUILDING.

NORTH GLASS FOR COOL LIGHT ALL YEAR

DIFFUSING SCREENS TO BOUNCE WINTER LIGHT THROUGHOUT ATRIUM

REFLECTIVE VENETIAN BLINDS

EXPOSED CONCRETE ABSORBS HEAT OF LIGHTS, PEOPLE & EQUIPMENT IN SUMMER.

office equipment. At night, as the outside temperature drops below that of the warm thermal mass, the building's ventilation system will flush the structure with cool air, a process that may continue until dawn.

In addition to the use of thermal mass and night ventilation, the building has a rock bed cooling system. Two containers, each filled with 660 tons of 1-inch diameter river rock, are set beneath the floor of the court. They are pre-cooled by blowing night air through the rocks. The coolness stored in the rock bed can be drawn on during the warm hours of the following day. The rock bed will provide 75 percent of the air conditioning required beyond that resulting from night ventilation. The rock bed can also be used to store heat during the winter; warm air is drawn into the rock bed during the afternoon and stored for use in the early morning hours.

Along the perimeter of the building, heat infiltration is regulated by the use of trellises on the south side and exterior shading on the east and west sides. The trellises act as an overhang, blocking summer sun but allowing the low winter rays to enter. Computer-controlled exterior shades on the east and west will lower to shade windows from direct sunlight and retract when the sun is off the window pane. Exterior shading prevents heat from entering, allowing for greater window area without adding to the cooling problem, and larger windows reduce the amount of artificial light needed in work areas.

Lighting in Site One will require half of the power required by a conventionally engineered office building, while providing better lighting conditions for office workers. Much of this savings is achieved by the use of individually controlled lighting for specific work areas. Where that type of lighting is not possible, in conference rooms for example, louvered flourescent fixtures will be used.

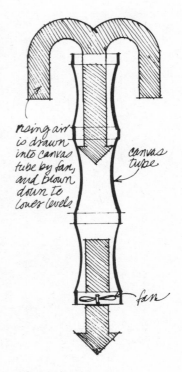

rising air is drawn into canvas tube by fan, and blown down to lower levels.

canvas tube

fan

CANVAS TUBE CIRCULATION SYSTEM

Site One will have a computer to monitor effects of various weather conditions. It will compensate for unusual conditions such as particularly hot days that are not followed by adequate night cooling. The computer will select the most efficient and economical use of the energy systems to maintain a comfortable temperature inside. It will also monitor the hourly energy consumption in the building and provide data with which to evaluate the effectiveness of the heating and cooling mix that was used.

Photo by John Lewis

Site One is now under construction in downtown Sacramento.

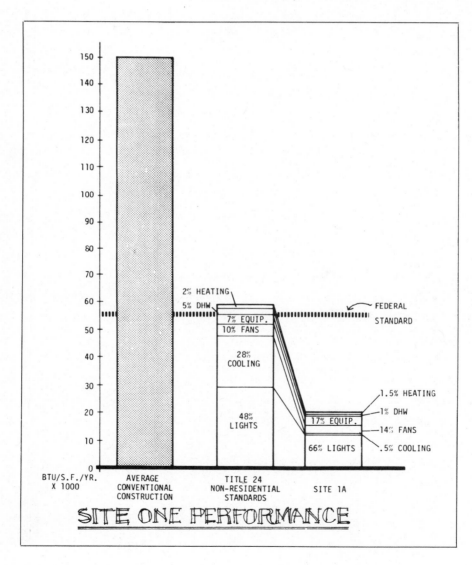

SITE ONE PERFORMANCE

Both Site One and the Terman Engineering Center are significant examples of architecture designed for a specific microclimate in order to both conserve energy and provide a pleasant, stimulating work environment. An added advantage of this approach is that no increase in construction costs is required. The projected construction cost of Site One is $59 per sq. ft., and is less than or comparable to that of similar new buildings.

Photo by John Lewis

45

Village Homes

The case studies in this book have focused mainly on cost-effective techniques for reducing energy consumption in residential and commercial structures. Often neglected, however, is the expenditure of energy and resources for the delivery of basic necessities such as food, energy, transportation, and waste disposal to an entire community.

There are many ways for newly built, planned communities to reduce a development's environmental impact while improving the quality of life of the residents. Energy-conscious land use planning like that used in Village Homes in Davis is the most important way.

In Davis, several innovative developers have resorted to energy-saving land-use planning to complement the city's conservation-oriented building code. One example is a 70-acre subdivision called Village Homes. The developers, Michael and Judy Corbett, have devised a land-use plan that conserves energy and resources while enhancing the sense of community shared by residents.

Houses in Village Homes are arranged in eight-unit clusters, with lot sizes that average 55 feet by 70 feet, approximately 4,000 to 5,000 sq. ft. apiece. Each lot is large enough for a private yard. Each cluster of eight homes shares one-third of an acre of land that is owned by the Village Homes Homeowners Association.

Owners of the common space mutually determine how it will be used; some have barbecue areas and lawns, while others have devoted the space to growing food. One advantage of shared responsibility for commonly owned land is that it increases the sense of community that characterizes the residents of Village Homes.

In Village Homes the automobile has a lower priority than people and open space. Roads are narrower than usual, 20 to 25 feet wide, rather than the standard 44 feet. The advantages are that it makes more land available for other uses, reduces the amount of asphalt and thus reduces reflected heat in the summer, and it is easier for trees to shade the streets. These factors make a considerable difference where the daytime

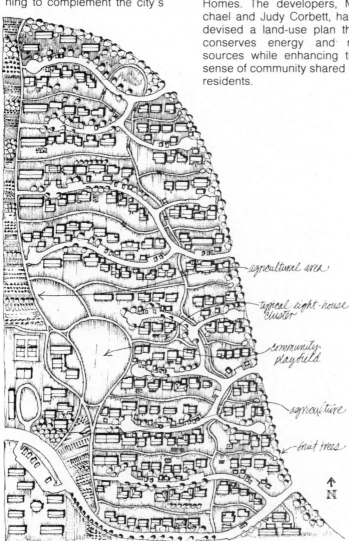

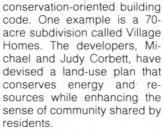

agricultural area

typical eight-house cluster

community playfield

agriculture

fruit trees

↑
N

PLAN VIEW of VILLAGE HOMES.

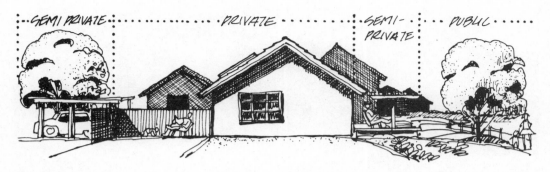

SEMI PRIVATE · · · · · · · · · · · · · · · PRIVATE · · · · · · · · · · SEMI-PRIVATE · · · · · PUBLIC · · · · ·

temperatures are about 100° F most of the summer. In addition, narrow streets cost less to build. In a typical development, the cost of streets ranges from $4,800 to $5,000 per house. In Village Homes the cost was approximately $4,000 per house, including parking bays, bike paths, drainage, and landscaping.

Long, narrow cul-de-sacs weave through the clusters, providing access but not dominating the landscape. Because neither the streets nor the cul-de-sacs are bordered by sidewalks, they appear to be narrow, winding roadways. Cul-de-sacs are long, averaging 400 feet; in a typical development they are approximately 200 feet in length but require more paved area. Minimal speed limits have been imposed throughout Village Homes. Off-the-street parking is available in parking bays owned by the Village Homeowners Association.

In addition to the common areas between the houses, greenbelts separate the clusters. Greenbelts are commonly owned and run north and south

through the community, perpendicular to the cul-de-sacs. Foot and bicycle paths go from cluster to cluster so that children can visit neighbors and play without crossing streets.

Throughout the greenbelts and around the clusters are small waterways filled with grass and stones. The waterways meander throughout the development and are used to collect rainwater from the lots and roadways. Around these areas are planted apple, apricot, walnut and pear trees. Nearly all of the rainfall is locally re-absorbed instead of being carried away through the traditional costly web of drains and pipes.

Maintaining the agricultural productivity of this land has been one of the Corbetts' goals. Nearly 12 acres of land have been set aside for small-scale, non-commercial use of agricultural land for crops, orchards, pasture, and vineyards. When the development is completed, the Corbetts expect that nearly 50 percent of the land in Village Homes will be under cultivation.

Work is beginning on a rec-

reation center that will include meeting rooms, an arts and crafts area, a day-care center, and areas for other activities. Soon a small commercial center will be added that will include a cooperative store, a small restaurant, an inn, and some small offices.

Even though not all the houses in Village Homes use solar energy, they all conform to the city's energy code. To date, nearly 100 homes have been constructed out of the 220 lots that are available. Fifty of them have solar hot water heaters, and all use some type of passive heating technique. All except two of the houses have conventional (mostly gas) back-up heating systems. And only one out of 10 houses has conventional air-conditioning. The Village Homes Architectural Review Board reviews all house designs and requires some uniformity in building materials. Also required is a design that will reduce the amount of energy used to heat and cool a home. The Architectural Review Board monitors all new construction to be certain that southfacing glass and so-

lar collectors are not shaded by another structure. Mike Corbett has built three-fourths of the houses himself; some he built on speculation and others were custom built for clients. Houses range in size from 900 to 2,500 sq. ft. and in price from $38,000 to $135,000. Ten percent of the Village Homes were built by their owners.

All the houses are aligned to take maximum advantage of the sun, and all homes have a south-facing exposure. Solar rights are guaranteed through the Covenants and Restrictions of the Homeowners Association so that no resident can block a neighbor's sun by planting tall trees. Deciduous trees can shade south-facing glass on specific dates during summer and fall. Roof-top solar panels cannot be shaded at any time of the year between 10 a.m. and 2 p.m.

Originally, the Corbetts had all eight houses in a cluster hooked up to a septic tank system with leach lines running through the cluster's common area. These plans were scrapped and all of the Village Homes are now connected to the city's sewer system. Corbett said that one important factor responsible for the development's success has been

the Davis branch of Guild Savings and Loan, which financed many of the houses.

Village Homes proves that resource conservation is feasible and practical in a private development. There has been little turnover of residents in the Village. As of spring 1978, only five houses had changed

hands and all were promptly resold. A recent study at the University of California at Davis compared energy consumption in the first unit of the Village to consumption in a similar neighborhood elsewhere in Davis. Village Homes is using 56 percent less total energy.

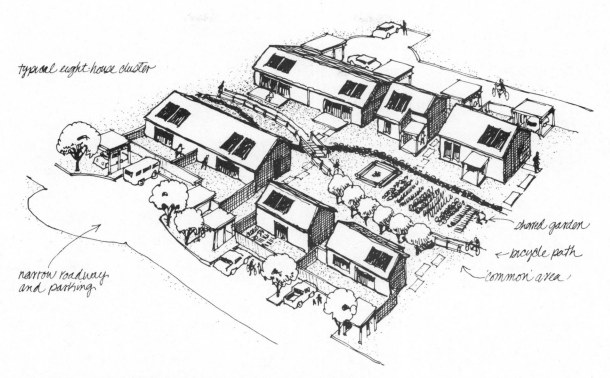

typical eight-house cluster

narrow roadway and parking

shared garden

bicycle path

'common' area

The obvious conclusion is that developers and residents can do well by doing good. Homeowners benefit from lower energy costs, a safer, more neighborly community, and self governance, while the developer does not have to pay for wide roads, sidewalks, and a storm sewer drainage system.

photo by Judy Corbett

III. INCREASING LOCAL INDEPENDENCE

Village Homes demonstrates that community-wide resource conservation is possible when the houses are designed and oriented to fit the climate and careful planning is done to reduce the adverse effects of urban encroachment. The financial success of the development indicates that an integral urban neighborhood combining local enterprise, food and energy production, and waste recycling in self-reliant communities is economically feasible.

However, most established communities do not have the same flexibility as a newly planned development. Residents must work with existing buildings, services, and conditions that were created long before the waste of energy and resources became evident. An obvious example is the network of highways that crisscross most urban areas, creating dependence on cars while excluding other forms of more efficient transportation. Similarly, the construction of centralized, high technology waste-treatment facilities prevents development of less costly and more desirable alternative methods.

This section deals with such alternatives as recycling solid wastes, treating wastewater in new ways, generating energy from renewable resources, and bringing food production closer to home.

COMMUNITY FOOD PRODUCTION

Land use planning at Village Homes enables half of the area to remain in food production. This keeps fertile, formerly agricultural land in use, brings food production closer to home, reduces food costs, reduces the use of energy to grow, process, and transport food, and results in a better quality of produce.

Localized food production is also possible in more urban, established communities. California has about 1000 community gardens in places ranging from the heart of downtown Los Angeles to small towns in the Central Valley. Increasingly, communities are turning unused space into productive, food-growing areas. These urban gardens provide crucial green space, a wildlife habitat, and they save money; a carefully tended 20 ft. by 20 ft. plot can save up to $400 a year in grocery bills.

At the Clementina Housing Project in San Francisco, about half a city block owned by the San Francisco Redevelopment Agency is serving as an urban garden. The land being gardened was originally cleared for another housing project, but it was never built. Originally, the Institute of Applied Ecology and San Francisco Community Garden Program cleared the site and laid out planting beds. Residents of the project then reorganized the plots to suit their own gardening patterns and techniques. The facilities were created entirely with recycled materials, and on one side there is a hill with a terraced garden.

Clementina Gardens has no site manager or organizer. The entire site is fenced, and the supervisor of the housing project has the key. More than 100 residents garden there, and plots are available for schoolchildren in the neighborhood. Each year residents hold an annual harvest and invite city officials and friends of the garden.

The largest community garden in California is at Leisure World, a retirement community of 20,000 people in El Toro, south of Los Angeles. Started more than 10 years ago by a former agriculture teacher, the garden now has 1,000 plots in two separate sites. Each site has a supervisor who sees that plots are maintained and that people who apply are assigned garden areas.

Individual plots are 10 feet by 20 feet, and the yearly fee is $10. The Leisure World recreation department furnishes water and tools. Horse manure and sawdust from the nearby stables are brought for fertilizer. Demand for garden plots is high; some people have been waiting more than two years. Gardening has become one of the primary recreational attractions of this retirement community.

WASTE DISPOSAL ALTERNATIVES

Communities of all sizes face the problem of waste disposal and treatment, which is becoming more serious as land and energy costs increase. Conventional waste management usually concentrates on treating rather than reducing or eliminating waste. That puts a burden on the homeowner, whose taxes pay for garbage collection, landfills, and sewage treatment.

A simple and economical method of reducing the production of waste is to stop it before it begins. This means consuming carefully, consuming less, and recycling. Recycling programs in many California communities are beginning to reduce solid waste.

In the case of wastewater, a state law passed in 1977 allows communities to establish local districts to oversee and upgrade waste treatment devices such as septic tanks and dry toilets. Where centralized sewers are already in use, biological processes can purify wastes, sharply reducing the cost of traditional methods of treatment. Waste reduction, management, and reuse strategies are often less costly than conventional methods. At the same time, they reduce consumption of resources, convert waste into a resource, and decrease the amount of energy and land needed for disposal.

53

Recycling

Recycling. For every person in California, one ton of solid waste is collected each year from residential, commercial and institutional sources. Strategies are being developed to reduce and reuse recoverable portions of this waste, reduce the amount of land needed for dumping sites, and conserve new materials needed for the production of goods.

Typical of California's recycling programs is one in Modesto, where more than 7,000 families recycle cans, glass, aluminum, cardboard, bottles and newspaper through the Ecology Action recycling program. That program combines all the elements of a sustainable solid waste resource policy: emphasis on reducing consumption, home separation of wastes, and the sale and

transfer of materials to be recovered for reuse. Thirty percent of the households in Modesto participate in this program through curbside pick-up and drop-off, and direct purchase outlets.

Cans, mixed glass, and newspaper are collected at the curbside pick-up routes. Recyclable materials are picked up twice a month, and curbside pick-up service is free. At the main drop-off center, cans, newspaper, cardboard, glass, magazines, and motor oil are accepted. The center is open 24 hours a day. Ecology Action will also buy glass, aluminum cans, and newsprint.

All the recyclable materials brought to the center are then processed for resale. Aluminum cans are shredded and sold. Newsprint is delivered to a local insulation manufacture and made into cellulose insulation. Glass is delivered to the Gallo glass furnace and melted down to be remade into wine bottles.

Local public awareness of the program is growing. The United States Environmental Protection Agency awarded a grant of $18,000 to Ecology Action to cover publicity costs

and to develop apartment routes. The insulation manufacturer has contributed money for door-hangers to advertise the curbside pick-up program. Participation has steadily increased, but Ecology Action still depends on an additional subsidy of employees funded by CETA (Comprehensive Employment Training Act).

Ecology Action has demonstrated that with concerted and cooperative effort, a local, independent recycling program can convert waste into a marketable resource.

Reusing Wastewater

At the Farallones Institute Rural Center in Occidental, greywater is used for subsurface irrigation. The greywater filtering system serves a small house containing a bath, bathroom sink, and a kitchen where milking equipment is washed daily. Total greywater output is 75 gallons per day. The house has been replumbed to carry greywater to a settling tank, sand filter, and holding tank. After the water is filtered, a siphon in the holding tank takes it to underground pipes in the garden, where it is used for irrigation. If the filter gets overloaded or clogged, a simple valve redirects greywater into a leachfield. Max Kroschel, designer of the installation, estimates that it requires one hour of maintenance a month and the sand filter must be cleaned every six months. As a do-it-yourself project, it can be built for about $300.

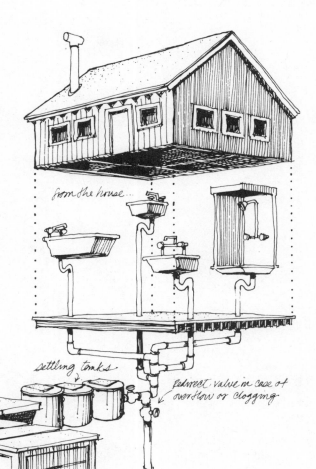

from the house...

redirect valve in case of overflow or clogging.

settling tanks

dosing siphon

holding tank

sand filter

to the leachfield

irrigation system

...to the garden.

The Mira Flores Housing Development in Tiburon uses a greywater system that carries wastewater from sinks, tubs, and laundry facilities to a holding and settling tank. Greywater is then filtered through a standard swimming pool filter and pumped back into the toilets for reuse.

Wastewater Management

Wastewater Management Districts. In 1976, the town of Stinson Beach, north of San Francisco, was authorized to form what is now termed an On-Site Waste Water Management District—a public agency that regulates, monitors, and inspects all local on-site waste disposal systems, primarily septic tanks. Unlike the other case studies in this book, this approach is neither a new technology nor a construction technique. However, it is a unique management method that qualified local public agencies throughout the state can use to improve existing on-site treatment facilities while avoiding the expense and adverse environmental effects of a centralized sewage treatment plant.

Like many other small towns, Stinson Beach was advised to build a sewage treatment facility by the local Water Quality Control Board, which claimed that the existing septic tanks and leachfields were inadequate and would eventually lead to deterioration of the local environment. The Stinson Beach County Water District, which is responsible for providing sewerage services, hired a consulting engineer to determine where and how this treatment facility could be built.

Because of the unique location of Stinson Beach on a sand spit surrounded by steep cliffs and the Golden Gate National Recreational Area, the proposed treatment system would have been expensive and would have adversely affected the local watershed. The town decided to keep the existing on-site septic tank-leachfield systems, but special state legislation gave the Stinson Beach Water District authority to regulate and inspect all on-site systems within its jurisdiction.

The water district monitors and inspects all septic tanks in the town and regularly tests the

56

quality of ground and surface water. In addition, the district has developed ordinances to prevent building over existing septic tanks, and all septic tanks must be accessible for regular inspections. New systems or system improvements must be of an acceptable design. The district has the power to evict persons for noncompliance with the standards established by state and local agencies. To date, this management district has successfully maintained water quality in the Stinson Beach-Bolinas Lagoon area.

The Stinson Beach Water District was used as the model for state legislation that enables certain public agencies in California to establish an on-site wastewater management zone. The management zone has the power to collect, treat, reclaim, and dispose of waste-

water without the use of sanitary sewers; to adopt and enforce regulations for the purpose of managing the on-site wastewater management zone, as long as the regulations do not conflict with those of the local health agency; to enforce all the zone's rules and regulations; and to charge for the correction of on-site system failures and for other services that the zone may offer.

An on-site management zone has several advantages, particularly for rural towns. It eliminates the need for and expense of centralized treatment facilities. A local, publically controlled entity can ensure adequate wastewater treatment and solve local wastewater treatment problems. Additionally, the management zone can provide technical information for septic tank and leachfield designs and maintenance techniques.

Converting Treated Sewage into Fertilizer

Converting Treated Sewage Into Fertilizer. Existing centralized sewage treatment plants are developing ways to turn waste residues into a marketable fertilizer. Vermiculturalist Jack Collier is experimenting with a method of using worms to digest sewage sludge at the San Jose sewage treatment plant. Sun-dried sludge is formed into windrows, watered, innoculated with worms, and covered with black plastic sheeting to retain heat and stabilize the temperature at approximately 74 degrees. In this warm environment, the worms rapidly break down the treated sewage into odorless, nutrient-rich castings. The worms also introduce beneficial bacteria that eliminate pathogens. Worms are then separated from the castings with a specially designed sorter.

The nitrogen/phosphorous/potassium ratio of the castings is high enough to be classified as fertilizer in California. Studies do not indicate yet if worms remove or concentrate heavy metals. Consequently, the sludge-derived castings can be sold as soil amendment for ornamentals and houseplants, but not as fertilizer for food crops.

Biological Treatment of Wastewater

Wastewater is purified as it passes through the aquacell.

Biological Treatment of Wastewater. Because of the high cost of conventional wastewater treatment facilities, there is a growing interest in techniques that use biological rather than chemical processes to purify wastewater. A process of this type has been developed by Solar AquaSystems, a research group in Encinitas. The treatment method is an attempt to recreate, in a closed environment, the continual purification of wastes by microorganisms and small invertebrates that occurs in nature.

The system relies on a series of enclosed, controlled ponds called "aquacells" through which wastewater is circulated and treated by aquatic plants and animals. Before entering an aquacell, wastewater receives primary treatment in settling tanks. Solids that settle out in this stage can be disposed of by anaerobic digestion or composting. Wastewater can then go directly into the cells, which are one- to two-acre ponds, 8 feet deep, covered by greenhouses to collect and retain heat. Each cell is aerated to maintain the correct oxygen balance.

In the cells, a dual process takes place. First, floating aquatic plants such as water hyacinths and duckweed remove greases and suspended solids, absorb toxic wastes, and most importantly, provide a suitable habitat for bacteria, and micro-invertebrates. The water hyacinths harvested from this treatment stage may be used as fuel. Water then enters a second treatment phase in which bacteria, aquatic plants, micro-invertebrates and filter-feeders consume nutrients and concentrate organics. Finally, the water is passed

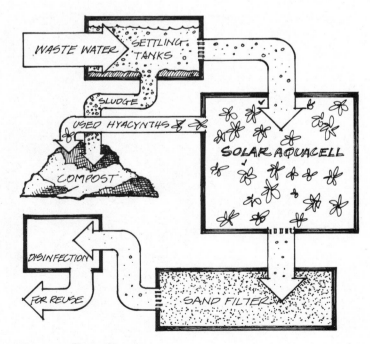

Process flow diagram

58

through sand filters and disinfected with either ozone or chlorine.

Various levels of treatment can be achieved, depending on how long water is retained in the cells. After two days in an aquacell, water quality is equivalent to secondary treatment. Tertiary quality water can be produced after six days in an aquacell. The potential exists to produce food and shellfish before sand filtration and disinfection.

The developers of this wastewater treatment technique point to its potential advantages: lower construction costs (40–60 percent less than traditional methods depending on the quality of treatment

desired), reduction of operating and energy costs, year-round operating efficiency due to solar greenhouse covers, natural reduction of pathogenic bacteria, and the poten-

tial harvest of fish, shrimp, and other useful byproducts—such as biomass for fuel, organic mulch, compost, or livestock feed. The facility uses about the same amount of space as conventional treatment plants, but it does require a temperate climate to work successfully.

The town of Hercules, near San Francisco, is constructing the first treatment plant using the Solar AquaSystems design. Construction started in 1978; completion is scheduled for late 1979. This will be the first full-scale treatment plant of this type in California and should provide valuable performance and cost data on biological wastewater treatment and reclamation.

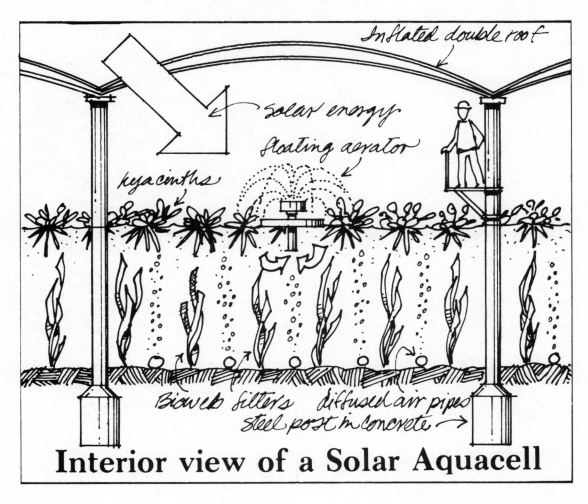

Interior view of a Solar Aquacell

ENERGY PRODUCTION FROM WIND AND BIOMASS

The projects in this book show how solar energy can be used to heat or cool buildings. Two equally important solar energy resources that have a variety of uses and play a critical role in the energy sustainability of communities are wind and biomass.

Over the past year, the use of these solar resources has expanded and the technologies for converting biomass and wind to useful energy are becoming increasingly refined. The projects mentioned here are appropriately scaled to the type and amount of energy required and are representative of the most sophisticated developments to date.

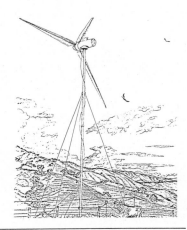

Biomass Energy

Biomass is organic matter: plants, trees, manure, and various agricultural and forestry wastes. These materials may be found in municipal refuse dumps, feedlots, building debris, tree prunings, or any place where organic material is grown, harvested, used, or deposited. Biomass can be converted to a clean and useful source of heat, steam, alcohol, methane, hydrogen, sulphur-free char, and synthetic fuels.

There are two primary bioconversion processes: microbiological and thermochemical. Microbiological processes rely on the action of microorganisms to decompose and convert organic matter into a more useful form. Examples of this process are fermentation used in the production of liquid fuels (alcohol, methanol, ethanol), and anaerobic digestion to produce methane. In microbiological processes, biomass can be used both as an energy source and as a source of organic material. After useful fuels are extracted through fermentation or anaerobic digestion, biomass can be used as a fertilizer, mulch, or in a variety of other ways.

Thermochemical processes rely on heat to convert biomass to fuel. These processes range from direct combustion (burning) to pyrolysis (chemical decompositon brought about

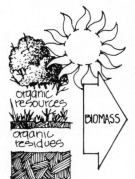

BIOCONVERSION PROCESSES / PRODUCTS

MICROBIOLOGICAL	THERMOCHEMICAL	PRODUCTS
methane fermentation		methane gas, fertilizer, ash
alcohol fermentation	distillation	ethanol, organic residue
	direct combustion	steam, high temp heat, ash
	pyrolysis	oil, gas, char
	hydro-carbonization	crude oil, SNG, fuel oil, ash

by the action of heat in the absence of oxygen.) Direct combustion can be used to generate electricity or steam for industrial processes. Pyrolysis, and a related conversion technique called gasification, produces fuels such as char, oil, and gas. The type of conversion process used to produce energy will vary with the availability and type of organic matter and the product or fuel desired.

California has a rich supply of biomass. Estimates of the total annual amount of available biomass residues are placed at more than 32 million tons. If all this could be converted to energy, it would supply over 10 percent of the state's current energy requirements, or virtually all electrical energy needs. Much of this biomass is being used as livestock feed, field mulch, soil conditioners, paper products, and other beneficial uses. For example, in California's sawmills, over half of the energy for space heating, process heat and electricity is supplied by the direct combustion of wood wastes. The amount of biomass wastes available for conversion to energy is approximately 22 million tons per year.

In California, new approaches to biomass conversion are being developed. In the agricultural community of Yuba City, a small corporation has built a prototype of gasification system that uses forest and agricultural residues and converts these wastes to energy. The system, developed by Biomass Fuel Conversion Associates (BFCA) requires sorted and prepared fuels. It can be used to replace natural gas and fuel oil; and can be adapted to convert biomass to electricity. The BFCA Gasifier is mobile, and can be used in different places depending on the availability of fuels. This increases the unit's flexibility; thereby, maximizing its efficiency and economy. This particular gasification system is locally fabricated and is assembled with standard components.

On a larger scale, the California State Legislature has allocated funds to build a biomass fuels system to heat state buildings in downtown Sacramento by producing a low-Btu gas that can be burned in existing heating and cooling plant boilers. Local organic wastes, such as landscape prunings, garden refuse, and agricultural and timber wastes will fuel these biomass gasifiers.

Unlike solar and wind energy, biomass can be converted to a hydrocarbon-based fuel that can directly replace fossil fuels. Turning biomass residues into fuel eliminates the cost of organic waste collection and disposal and provides a local, renewable, and low-cost fuel source.

The State of California's biomass fuel system will adjoin the existing heating and cooling plant.

61

Wind Energy

photo by Ken Smith

Recent developments in Wind Energy Conversion Systems (WECS) are expected to make wind a more widely used energy alternative in California. In 1978, the legislature determined that wind energy systems will be classified as a solar technology, making them eligible for the 55 percent tax credit. This should stimulate the market for small scale (3 to 10 kilowatts output) wind systems, which are still very expensive for the individual user. Wind systems for commercial use will get a 25 percent tax credit. Guidelines to qualify these systems for the tax credit have been developed by the California Energy Commission and the Office of Appropriate Technology.

In some parts of California, persons generating electricity from wind can sell excess power to a utility and buy power back when they need it. Southern California Edison has such a program for both residential and small commercial customers with wind systems. By eliminating the cost of storage batteries, the total cost of wind energy is reduced. But, because the utility becomes the storage, they buy power at a lower cost than they will sell it.

The appropriateness of a wind energy system is related to a variety of factors which include site, available wind, and local climatic conditions. On the basis of both cost and maintenance, wind machines in the 20 to 40 Kw range are becoming increasingly attractive. These wind machines can serve a cluster of houses or a small neighborhood, reducing the cost of the unit by sharing of storage and generation facilities.

U.S. Windpower Associates will be building a series of 40-kilowatt wind turbines in California and has signed a contract with the State Department of Water Resources to sell wind-generated electricity from the nation's first "wind farm." The state agency will purchase 2.5 million kilowatt-hours of electricity from June of 1981 to April of 1983 at the rate of 3.5 cents a kilowatt-hour. The company will initially build 20 wind turbines at a state-owned site near Pacheco Pass, south of San Francisco. The estimated cost of this privately financed effort is $19 million.

California's first large wind generation system (3,000 kw) is being planned by Southern California Edison. The utility is purchasing a test prototype from Wind Power Products of Seattle, and installing it at a test site in the San Gorgonio Pass near Palm Springs. The projected cost is about $2 million. Testing will start in 1979.

New efforts such as the ones briefly mentioned here are becoming increasingly common as more people and utilities recognize the potential of these renewable energy sources. Although wind energy systems and biomass conversion techniques have not received as much private or governmental support as other solar technologies, they are beginning to make significant contributions to the use of renewable resources in communities.

FINANCING ALTERNATIVE SYSTEMS

All the techniques described in this book for residential and commercial buildings require money. Consequently, their success or failure in the marketplace depends on how willing lenders are to finance them. In the past, lenders have seen energy-conserving designs or alternative energy systems as high-risk investments. They have been convinced neither that the technologies work nor that there is a market for buildings that incorporate them. This attitude is changing. As the examples in this book indicate, lenders' experience in financing alternative energy systems is growing, and institutions are becoming more willing to offer financing provided that the technologies are well-designed, efficient, and offer an adequate rate of return.

For solar systems, lenders make two types of loans: home improvement loans, generally for domestic hot water heaters, and new home loans. The two are treated quite differently. According to a recently completed survey of commercial banks and savings and loan associations done by the State of California's Solar Business Office, only 12 percent of the respondents would include solar systems in new mortgages, while 46 percent favored solar home improvement loans. This suggests that lenders' attitudes are tied to the potential rate of return on a loan, not to the new technology; home improvement loan packages offer the lender an opportunity to obtain a new-loan origination fee and to renegotiate or extend the terms of the existing mortgage.

Many banks and savings and loans have started or are exploring solar home improvement loans, primarily for systems that heat domestic water. There are several types of solar home improvement loans being offered. San Diego Federal Savings and Loan Association has pioneered a type of solar loan that involves renegotiating the first deed of trust to incorporate the solar loan. The solar loan is given at the current interest rate, and the original mortgage is extended so that the new loan can be incorporated without an increase in the consumer's monthly payment. Consumers pay a loan-origination fee, and mortgages are limited to thirty years.

This approach is particularly appealing where conventional fuels are expensive and borrowers will see an immediate return on their investment in the form of lowered utility bills, improved cash flow, and credits on state and federal taxes. One disadvantage of the arrangement is that the lender does not disburse funds until the installation is completed and operating. This procedure ensures consumer satisfaction and the safety of the loan, but it also delays the contractor's payment.

A factor that will limit adoption of this program is that the solar loan can be rolled only into an existing loan held by the lender and that has not been sold in the secondary mortgage market. Once in the secondary pool, the loan cannot be reclaimed to be renegotiated for a solar retrofit. Nonetheless, according to the Solar Business Office's survey, 64 percent of the respondents preferred this type of financing scheme. For a copy of the survey and a list of lenders with solar programs in California, write to the Solar Business Office, 921 Tenth Street, Sacramento, California 95814.

Other lenders such as Home Federal Savings and Loan offer a reduced interest rate for energy-saving devices. Home Federal offers up to a 2 percent discount for solar domestic water heaters and water-saving devices, and has hired an energy consultant to work with its customers.

Although Bank of America does not have a special incentive program for solar home improvement loans, it has hired Berkeley Solar Group, a respected company of solar engineers and consultants, to evaluate forty systems for performance and reliability. Although Bank of America will not necessarily deny a loan based on evaluation, it will ensure the security of its loan by informing the borrower when the consultant believes the system is inadequate.

In the case of a new mortgage for a solar home, lenders' doubts about the soundness of the technology come into play. Some lenders will provide money only when they have ensured the security of the loan. For example, they will not lend money for a system that does not meet local building codes. If the code requires a conventional fuel back-up, so does the lender. Large lenders may hire an architect or engineer to review plans for solar homes. Lenders are also concerned with resale value; they are likely to turn down unconventional designs with limited appeal.

One last factor that can make a solar loan particularly attractive to both borrower and lender is a solar tax credit. The State of California and the federal government offer tax credits as incentives to use both passive and active solar energy systems. The credits are subtracted from the tax an individual owes. The California credit is 55 percent of the cost of a residential solar system (or not more than $3000). The total cost of the system cannot exceed $6000, and the credit may be carried over from year to year. California also has a 25 percent solar tax credit for multifamily residences and for commercial buildings. Conservation measures that are associated with installing the system are also eligible for the tax credit. Wind energy producers, solar industrial process heat and solar production of electricity are all eligible for the 55 percent credit. For a copy of the guidelines and criteria for the California Solar Energy Tax Credit, write to the California Energy Commission, 1111 Howe Avenue, Sacramento, California 95825.

The federal government offers tax credits both for solar installations and for conservation measures. The solar credit is for 30 percent of the first $2000 of an installation's cost and 20 percent of the cost for the remaining $2000 to $10,000; the total credit may not exceed $2200. The conservation credit allows the taxpayer to subtract up to 15 percent of the cost of conservation measures; the ceiling is $300. Federal regulations do not allow for a commercial solar tax credit, but unlike the California regulations, they do allow geothermal energy producers to use the credits.

California's lenders are becoming more aware of the value and long-term economy of energy-conserving and alternative energy systems. Consumers should become familiar with both State and Federal solar tax credits and then find lenders that have attractive loan policies. By working closely with financial institutions, prospective builders and homeowners will ensure that lenders understand the economic and other merits of incorporating resource-conserving designs and techniques into buildings. As lenders better understand these techniques, they will be more likely to incorporate them into construction and mortgage loans.

APPENDIX

HOLLANDALE DAIRY
Designer: Bill Wilson, Modesto Junior College.
Coordinator: Reza Vossoughi, Conservation Engineer, Modesto Pacific Gas and Electric.
Owner: Ted Dykzeul, Hollandale Dairy.
Contractor: No contractor. Project was sponsored by Hollandale Dairy, Energy Systems Inc., Modesto Junior College, Pacific Gas and Electric, Turlock Irrigation District, Wolverine Solar, the Isotherm Co., and the University of California Extension Service.
Completion date: February 1978.
Size: Six flat-plate collectors manufactured by Energy Systems, Inc.; total surface area 102 sq. ft.
Location: Oakdale, California.
Type of system: Active hot water heating system for equipment sterilization and cow priming.
Storage volume: 360 gallons, 120 of which are connected to supplementary conventional water heater.
Type of storage: Three 120-gallon tanks.
Insulation: Water tanks were thoroughly insulated.
Cost: Similar system would cost an estimated $5,050; this project cost less because it was a prototype, so companies involved donated some materials and labor.
Financing: None.
Back-up systems: 120-gallon gas-fired water heater.
Estimated conventional fuel savings: $704 per year in electricity, or about 41 percent of previous bill.
Construction time: Five weeks.
Expected system life: More than 20 years.
Payback time: 7.17 years ($5,050 divided by $704).
Average insolation: 592 KBtu/1 sq. ft./month, horizontal surface.

LA FRANCHI DAIRY HEAT RECOVERY SYSTEM
System: Heat is recovered from the cooling compressor used to chill and refrigerate the milk; then heat is used to heat the milking parlor washdown water. Eighty-degree water for washing the cows is drawn off from a midpoint in the heat exchanger tank. The system heats the rest of the water to 142 degrees; then the water enters a conventional water heater, which raises the water temperature to 170 degrees. The system uses waste heat from two five-horsepower water-cooled compressors.
Equipment: Heat exchanger manufactured by Dairy Equipment Company, P.O. Box 1289, Madison, Wisconsin 53701.
Electrical use: Before installation—total, 7714 kwhe; average daily, 118 kwhe; daily cost, $4.47; after installation—total, 1876 kwhe, average daily, 20 kwhe; daily cost, $0.81.
Costs: Equipment—$1,599; installation—$181 (can go up to $900 for this unit depending on distance between cooling compressor and water heaters).
Savings: $1,337.36 yearly.
For further information: "Heat Exchangers Convert Btu's to $$$," David Halsey, *Dairy Herd Management,* January 1978, available from Miller Publishing Company, 2501 Wayzata Blvd., Minneapolis, Minnesota 55440.

HEWLETT-PACKARD SYSTEM

Designer: Martin McPhee, Maintenance Chief.
Owner: Hewlett-Packard Corporation, Automatic Measurement Division.
Contractor: Constructed by company personnel.
Completion date: Spring 1974.
Size: 165,000 sq. ft. in building; 10,000 sq. ft. of flat-plate collectors.
Location: Sunnyvale.
Type of system: Active hot water system—terminal reheat system. Solar heated water preheats boiler water, reheats air from the air conditioner, and heats domestic water.
Storage volume: 14,000 gallons.
Type of storage: Underground tank.
Cost: $30,000 exclusive of labor; cost/sq. ft.—$3 per sq. ft. of collector.
Financing: None.
Back-up systems: Not applicable; preheats water for existing space heating and air conditioning system.
Estimated conventional fuel savings: $2,000 per month, or 65 percent of total gas bill.
Construction time: 1,120 hours.
Expected system life: 20 years.
Payback time: About 1.5 years, exclusive of labor costs.

RENAULT AND HANDLEY SOLAR BUILDINGS

Two solar buildings are part of this complex. All information is for Trombe wall, or for all of building #1, as specified.
Architect: Harry Whitehouse, Pacific Sun, 540 Santa Cruz Avenue, Menlo Park, California 94025.
Owner: Renault and Handley, Inc., a development firm.
Contractor: Buffalo's Mechanical, Mountain View.
Completion date: October 1978.
Size: Building #1—45,000 sq. ft.; Trombe wall—1,200 sq. ft.
Location: Santa Clara.
Climatic variables: See table.
Type of system: Passive—Trombe wall, air circulation by convection and by fan.
Storage volume: 600 cu. ft. in Trombe wall.
Type of storage: Concrete structure.
Insulation: Ceiling—R-19; walls—vary according to location because lessees finish interiors; varies from about R-6 to about R-10.
Cost: For total building—$1.25 million; for Trombe wall—$22,000. Cost/sq. ft. for Trombe wall—$18.33 (could have been cheaper had builder used plastic instead of glass for glazing).
Financing: For active solar systems, the cost was shared equally by Department of Energy and by the developers. For balance of building, owners financed through Wells Fargo.
Back-up system: Gas furnace.
Estimated conventional fuel savings: About $600 per year at 1978 natural gas prices. System is now being monitored by Pacific Sun for more accurate savings.
Construction time: One year; the builder was in no hurry, so timing was controlled by manipulations in leasing arrangements because builder was tailoring to lessees' requirements.
Expected system life: Life of the warehouse.

Payback time: 17 years for total system including Trombe wall. Payback time is based on assumption that the price of natural gas will escalate at only 15 percent per year.

Other important factors: The unique aspect of this system is that building it required only a simple adaptation of a common construction technique. The Trombe wall did not add much to construction cost because it was the same thickness as the other structural walls. Cost and performance of Trombe walls will vary for other structures depending on design.

INTEGRATING A SOLARIUM INTO NEW HOME

Designer: Peter Calthorpe, 2040 Larkin, San Francisco, California 94109.
Owner: Confidential.
Contractor: Farallones Institute, 15290 Coleman Valley Road, Occidental, California 95465.
Completion date: Fall 1976.
Size: 1200 sq. ft. interior; 300 sq. ft. solarium.
Location: Occidental.
Climatic variables: See table.
Type of system: Hybrid. Greenhouse with a 65-sq. ft. attached flat-plate air collector heats air in winter, provides ventilation in summer; heat is stored in slab floor and the rockbed underneath; air circulates partly by convection and partly by fan.
Storage volume: 742 cu. ft. of ¾-inch rock.
Type of storage: Rockbed.
Insulation: Ceiling — R-19, 6-inch fiberglass; walls — R-11, 3½-inch fiberglass; perimeter — 1½-inch styrofoam.
Cost: Total — $40,000 — no finished floor in living room and no kitchen cabinets. Estimated $43,000 finished. Cost/sq. ft — $28.67.
Financing: None.
Back-up systems: Fireplace — woodburner, required by county.
Estimated savings: About 28,000,000 Btus/year.
Construction time: Six months.
Expected system life: Life of the house.
Other information: Solarium provides both heat and light because it is built as an integral part of the house. It also provides food and serves as a buffer between house and outdoors because it contains the main entry.

ATTACHED SOLAR GREENHOUSE (retrofit)

Designer: Lynn Nelson of the Habitat Center with Jeff Reiss, John Burton; 573 Mission Street, San Francisco, California 94105.
Owner: Jeff Reiss.
Contractor: Habitat Center and Jeff Reiss; not union labor.
Completion date: October 1978.
Size: 21 feet by 7.5 feet; 200 sq. ft., including alcove.
Location: Sacramento.
Climatic variables: See table.
Type of system: Passive — greenhouse with 12 water-filled 55-gallon drums absorb heat; air circulates through house by convection. In summer, overhang shades greenhouse to reduce heat gain, and opened vents provide circulation. Provides 64 percent of heating for house in winter and all of summer cooling; estimate based on heat loss and solar gain.

Storage volume: 660 gallons of water.

Type of storage: Water-filled drums.

Insulation: (greenhouse only) Ceiling—R-20; walls—west wall R-20; perimeter—none.

South-facing glazing: 120 sq. ft.

Cost: Total—$16,000; cost/sq. ft.—$8.

Back-up systems: Because this is a retrofit, the house has a complete heating system already built in. The greenhouse provides 64 percent of the house's heat in winter.

Estimated savings: 36 percent of previous use for heating. Cooling system supplies 100% of owner's needs.

Construction time: Constructed during weekend greenhouse workshop.

Expected system life: Life of the house.

THE GREENSTEIN RESIDENCE

Architect: Rob Quigley, Gluth and Quigley Associates, 662 State Street, San Diego, California 92101.

Owner: The Sherman Greenstein family.

Contractor: Bottom Line of Santa Monica, a firm owned and operated by two architects.

Completion date: Winter, 1978.

Location: Woodland Hills, northwest of Los Angeles.

Climatic variables: See table.

Type of system: Hybrid—house oriented around belvedere, which acts as thermal chimney for ventilation. Space is heated by eight-by-twelve-foot solar box of water-filled 55-gallon drums. Fan on conventional furnace circulates air. Solar hot water heating.

Type of storage: Water columns and rockbed.

Storage volume: Rockbed—672 cu. ft.; water volume—528 cu. ft.

Insulation: Ceiling—partly R-19 and partly R-30; walls—R-11, windows are double-glazed; floors—R-11.

Cost: Total—$73,600; cost/sq. ft.—$46; $46 is competitive.

Financing: California Federal Savings.

Back-up systems: Gas furnace and conventional water heater, required by building codes and lender but also integral to design.

Estimated savings: The owner estimates that solar water heater supplies 100% of hot water needs from April to October and 50% during the winter months. The design of the house cuts heating needs by 80% and the cooling system supplies 100% of their cooling needs.

Construction time: One year.

Expected system life: Life of the house.

TOM SMITH RESIDENCE

Architect: Lee Porter Butler, Ekose'a, 573 Mission Street, San Francisco, California 94105. John Hofacre, Bruce Maeda and Carl Gustafson were also significant contributors to the concept and design of this house.

Owner: Tom Smith, P.O. Box 2356, Olympic Valley, California 95730.

Contractor: Robert Charbonneau.

Completion date: February 1, 1978.

Location: North shore, Lake Tahoe.

Type of system: Passive—uses thermal envelope composed of attached greenhouse on south and air passages in the roof, the north wall, and under the house.

Storage volume: 2160 sq. ft. backfill, 544 sq. ft. concrete.
Storage type: Backfill in foundation, concrete foundation walls.
Insulation: Air space—R-19 in roof and north wall and walls of greenhouse; living area—R-13 in roof and north wall.
Cost: Total—$53,258; cost/sq. ft.—$30; average in area—$35-$37 sq. ft.
Financing: Conventional mortgage by Bank of America.
Back-up systems: Wood stove in living room; not required by code or financial backer, but designed by architect to remove chill on severe winter days.
Estimated savings: The owner estimates that the savings for heating range between 70 and 80%. With an efficient wood-burning stove, savings could be as high as 100%. During the winter of 1979, the owner used only two-thirds of a cord of wood.
Construction time: Four months.
Expected system life: Life of the house.
For further information: *THE ENERGY-PRODUCING HOUSE   The How-To Handbook/Case Study,* Tom Smith and Lee Porter Butler, $18.95 from Tom Smith, P.O. Box 2356, Olympic Valley, California 95730—includes working drawings. Also *EKOSE'A HOMES: Preliminary Planning Package.* Ekose'a, $24.95 from Ekose'a, 573 Mission Street, San Francisco, California 94105—ten preliminary designs with floor plans, elevations, and sections.

DAVE SMITH RESIDENCE

Architect: Brent Smith, 3921 Dawn Drive, Loomis, California 95650.
Owner: Dave Smith
Contractor: Dave Smith, Creative Enterprises, 5341 Angelina, Carmichael, California 95608.
Completion date: October 1978.
Size: 2,500 sq. ft.
Location: Carmichael.
Climatic variables: See table.
Type of system: Hybrid—80 sq. ft. of greenhouse for each unit provides space heating, with fans for air circulation. Exposed slab floor is thermal storage mass. Wood stove and "chill-chaser" radiator system provide supplementary space heating. Insulating shutters and exterior shading regulate heat gain and loss. Solar collector system heats water. Gas-fired conventional water heater is a back-up for both space and domestic water.
Storage volume: 1,066 gallons of water in hot water tank; slab floor is mass for passive thermal storage.
Type of storage: Exposed aggregate slab floor.
Insulation: Ceiling—R-30; walls—R-19; perimeter—R-19.
Cost: Estimated cost would be $80,000; the actual cost was less because Dave Smith is a contractor, so his cost didn't include profit or labor. Cost/sq. ft.—estimated $32, same caveat as above. Average cost in the area—$35/sq. ft.
Financing: World Savings and Loan in Sacramento.
Back-up systems: Wood stove and "chill-chaser" radiator system, along with gas-fired conventional hot water heater.
Construction time: Because the house was owner-built in the owner's spare time, the construction of this house would not be representative of the time it would take to build a similar house.

HARMON RESIDENCE

Designer: Jim Harmon, Peter Hansen.
Owner: Jim Harmon.
Contractor: Owner-built.
Completion date: October 1978
Construction time: Approximately two years; done in spare time.
Size: 1,000 sq. ft.
Location: El Centro, 25 miles west of Calexico near Mexican border.
Climatic variables: See table.
Type of System: Passive—please refer to text for complete description of systems.
Type of storage: Mass of house absorbs and stores heat.
Insulation: Ceiling—air path and multiple layers of insulation, minimum of R-30 in narrowest portion; walls—minimum of R-19, maximum of R-50. Earth berms add additional insulation on north, west, and south sides.
Cost: Total—$22,000; cost/sq. ft.—$22.
Financing: None.
Back-up systems: 30-inch exhaust fan in the tower for increased air circulation, and pads and fan over air intake pipes.
Expected system life: Life of house.
Estimated conventional fuel savings: Not possible to estimate.

HODAM RESIDENCE

Designer: Bob Hodam.
Owner: Bob Hodam.
Contractor: Bob Rutherford, Buffalo Builders, P.O. Box 161616, Sacramento, California 95808.
Completion date: October 1977.
Size: 2,800 sq. ft.
Location: Sierra Nevada foothills near Placerville.
Climatic variables: See table.
Type of system: Hybrid—a greenhouse provides heat in the winter, northerly windows provide cooling; air circulates by convection; a reverse-cycle heat pump provides supplementary heating and cooling; a conventional electrical system heats the water for the reduced-flow system. This house's principal distinguishing feature is its integration of food production, waste recycling, and water conservation into the household.
Type of storage: Mass of house.
Insulation: Ceiling—R-30; walls—R-19; perimeter—none.
Cost: Total—$90,000; cost/sq. ft.—$33.
Financing: Combination of mortgages from Eldorado Savings and Loan and Golden One, the state employees' credit union.
Back-up system: Reverse cycle heat pump for both heating and cooling, required by lenders.
Estimated conventional fuel savings: 50 percent of a conventional house's use.
Construction time: Six months.
Estimated system life: Life of the house.
Other important systems: Water—flow restrictors, low-volume flush toilets, and water conserving showerheads throughout, saving not only water but electricity for heating the water; separate discharge lines for kitchen sink, bathroom shower and sink and laundry, so greywater may be stored and reused in periods of drought. Food—the owner is experimenting with hydroponics in the greenhouse and has plans to try aquaculture in greenhouse, too.

72

TERMAN ENGINEERING CENTER

Architect: Harry Weese and Associates, 10 West Hubbard, Chicago, Illinois.
Owner: Stanford University.
Contractor: Lathrop Construction of San Francisco; the mechanical engineers were Guttmann and MacRitchie, also of San Francisco.
Completion date: September 1977.
Size: 151,959 sq. ft.; 104,202 sq. ft. net assignable area (excludes areas such as store rooms not affected by heating requirements).
Location: Palo Alto.
Climatic variables: See table.
Type of system: The building encircles an inner core. The structure around the core is heated and cooled using passive systems such as louvered shutters, outside exterior doors, louvered doors, baseboard hot water convectors for winter heating, and skylights. Conventional heating system regulates conditions in the inner core, which contains laboratories, the library, and the auditorium.
Insulation: Ceiling—R-10; walls—R-factor varies considerably according to location; perimeter—concrete basement and first floor not insulated because concrete serves as thermal mass.
Cost: Total—$7,478,545; cost/sq. ft.—$49.81 (considerably under the projected $59); average in area—$70 to $80 for building of comparable size.
Financing: None. The building was a gift from William Hewlett and David Packard.
Back-up systems: Conventionally fueled central heating and cooling plant on Stanford campus. Back-up not required by code, but by design.
Estimated conventional fuel savings: 50 percent of heating bill of a comparable building.
Construction time: 17 months.
Expected system life: Life of the building, or about a hundred years.

SITE ONE

Architect: Office of the State Architect; State Architect—Sim Van der Ryn; Project designers—Peter Calthorpe, Bruce Corson, Scott Matthews.
Owner: State of California.
Users: The Department of Developmental Services and the Department of Mental Health.
Contractor: Continental Heller.
Completion date: Summer 1980.
Size: 267,000 sq. ft.
Location: Eighth and P Streets, Sacramento.
Climatic variables: See table.
Design features: 150 x 144-ft. four-story roofed atrium. Court daylit by roof skylights with operable shading (passive solar heater in winter, precooled by night ventilation in summer). Court connects building entrance to elevators, stairs, and balconies at each floor. All work spaces within 40 feet of daylit court or exterior window. Work spaces have 10½-foot ceilings, exposed structure, daylit at perimeter, indirect ambient lighting combined with individual task lighting. Exterior operable automatic sun shades on east and west-facing walls, fixed trellis on south. Rockbed thermal storage system augments use of night ventilation, pre-cooling of building structure, and efficient, computer-controlled mechanical system. Solar domestic hot water system. Exterior balconies accessible from each office work area.

Performance: 19,900 net Btu/sq.ft./year energy consumption, less than 40 percent of consumption of typical building built to California's nonresidential energy design code (Title 24). Computer controlled operation of the building's mechanical system will reduce demand and time-of-day electrical power charges, as well as virtually eliminate additional demand on the state government's central plant for additional peak cooling capacity, thus substantially reducing operating costs.

Cost: Total project cost—$19,616,000; construction cost—$16,629,500; cost/sq. ft.—$60.

Financing: State appropriation.

Estimated conventional fuel savings: $77,000 projected first year savings (1978 dollars) over building conforming to Title 24.

Cost/benefit: Site One will save taxpayers $684,000 (1978 dollars) during its economic life over a Title 24 building, and $3.1 million over a conventionally constructed building.

VILLAGE HOMES

Developers: Michael and Judy Corbett, Davis, California.

Size: 70 acres.

Location: Davis

Climatic variables: See table.

Provisions of the development: Houses arranged in eight-unit clusters; lots average 50 × 70 feet; each cluster owns about a third of an acre in a common area. Roads vary from 20 to 24 feet in width, reducing both cost and amount of land pre-empted by cars. Long cul-de-sacs of 400 feet (usual conventional length is 200 feet) provide access to houses but protect privacy. Greenbelts separate housing clusters and provide foot and bicycle paths. Meandering waterways collect drainage water and allow it to percolate and remain on the site. Developers have set aside 12 acres of land for cultivation by residents. Of the hundred homes that have been built on the site, 50 have solar hot water heaters; all use passive heating techniques; 2 have active space-heating systems; 7 have hybrid space-heating systems. All but two of the homes have conventional back-up heating systems. The Convenants and Restrictions of the Homeowners Association have guaranteed owners' solar rights from 10 AM to 2 PM. All houses are connected to the city's sewer system. Turnover in the development is low, and those houses that have changed hands were resold promptly. A study done by a University of California at Davis graduate student indicates that the development uses only 57 percent of the energy used by a similar neighborhood that was built to conform to the city's energy conservation code.

Costs: Houses—houses vary in size, and cost from $38,000 to $125,000 to build. All have southern exposure. Streets—approximately $4,800 to $5,000 per home, perhaps $100 to $200 less than in a normal subdivison, but that money also pays for bike paths, natural drainage, and landscaping, which is unusual.

For further information: "Low Energy Consuming Communities, Implications for Public Policy," Jan Hamrin, doctoral dissertation, University of California at Davis, March 1978, available through *Dissertation Abstracts. Village Homes Solar House Designs:* Bainbridge, Corbett, and Hofacre; to be published by Rodale Press in Spring 1979.

CLIMATE AND SOLAR DESIGN VARIABLES

| STRUCTURE | LOCATION | TEMPERATURE | | DEGREE DAYS HEATING 65°F BASE | DEGREE DAYS COOLING 65°F BASE | DESIGN TEMPERATURES (°F) | | INSOLATION KBTU/F₂/Mo. HORIZONTAL RADIATION |
		AVERAGE DAILY MAX / MIN	ANNUAL EXTREME MAX / MIN			SUMMER DRY BULB 0.2%	WINTER 0.2%	
RENAULT & HANDLEY BUILDING	SANTA CLARA	67/49	104/21	3566	420	88	34	579
REISS/HABITAT CENTER GREENHOUSE	SACRAMENTO	73/47	115/20	2843	1159	98	36	581
CALTHORPE DESIGNED HOUSE	OCCIDENTAL	68/46	106/23	3413	204	93	32	576
GREENSTEIN RESIDENCE	WOODLAND HILLS	69/54	110/30	1800	1310	88	32	559
DAVE SMITH RESIDENCE	SACRAMENTO	73/47	115/20	2843	1159	98	36	581
TOM SMITH RESIDENCE	TAHOE CITY	56/29	94/-15	8162	32	77	11	566
HARMON RESIDENCE	EL CENTRO	88/58	115/24	1010	4195	107	31	685
HODAM RESIDENCE	PLACERVILLE	73/47	110/15	4161	731	96	29	566
TERMAN CENTER	PALO ALTO	67/49	104/21	2969	236	85	34	579
SITE ONE	SACRAMENTO	73/47	115/20	2843	1159	98	36	581

GLOSSARY

ABSORBER PLATE—See flat-plate collector.

ACTIVE SOLAR SYSTEM—A system that uses mechanical devices and an external energy source, in addition to solar energy, to collect, store, and distribute thermal (heat) energy.

ANAEROBIC DIGESTION—The controlled decay of organic material in an air-tight container. Biogas results from this decomposition without oxygen.

BTU or BRITISH THERMAL UNIT—A measurement of energy representing the amount of heat needed to raise 1 pound of water by 1 degree Fahrenheit. About the same amount of energy released by a single lit match.

CLERESTORY—A vertical window placed near the ceiling that is used for light, ventilation, and to collect heat.

CONVECTION, NATURAL—Heat transfer through a medium such as air or water by currents that result from the rising of lighter, warm air and the sinking of heavier, cool air.

COST EFFECTIVE—Economical in terms of tangible benefits produced by money spent.

DEGREE DAY COOLING—A measure used to evaluate a location's summer cooling requirements. Each degree that the mean daily temperature is *above* 65 degrees F is called a cooling degree-day. The monthly value of cooling degree days is the sum of the degree-days accumulated for all days in the month. Cooling degree days are not particularly useful in the design of a building's cooling system since they do not include relative humidity effects, and because 65 degrees is well below the usual 78-degree cooling design temperature.

DEGREE DAY HEATING—A measure used to determine heat requirements. For any one day when the mean temperature is less than 65 degrees F, there exist as many degree days as there are Fahrenheit degrees difference in temperature between the mean temperature for the day and 65 degrees F. The sum of the degree days constitutes the annual degree day heating requirement.

DRY TOILETS—A waterless toilet (e.g., composting, humus, incinerating).

FLAT PLATE COLLECTORS—An assembly containing a panel (absorber plate) of metal or other suitable material, usually a flat black color on its sun side, that absorbs sunlight. This panel is usually in an insulated box with a covering of glass or plastic to retard heat loss. The sun converted to heat in this way is then carried by a circulating medium for use elsewhere.

GLAZING—A glass covering. This term can refer to a fiberglass covering, as well.

GREYWATER—Most easily defined by what it is not: it is not toilet wastewater ("blackwater"). Greywater comes from all other fixtures in the household.

HEAT EXCHANGER—A device for transferring heat from one medium to another, such as an old fashioned radiator that uses hot water to heat air. It can be as simple as a coiled piece of tubing immersed in a fluid storage tank through which collector heat passes.

HEAT GAIN—An increase in the amount of heat contained in a space, resulting from direct solar radiation and the heat given off by people, lights, equipment, machinery, and other sources.

HEAT PUMP—A heat pump is a combination heating and cooling unit that operates like a normal air conditioner in summer and in winter operates in reverse, blowing warm air indoors while pushing cool air outdoors.

HEAT SINK—A body which is capable of accepting and storing heat and, therefore, may also act as a heat source.

HYBRID SOLAR SYSTEM—A system that combines passive solar collection with active transport of heat to an isolated storage system.

INFILTRATION—The uncontrolled flow of air into a building through cracks, openings, doors, or other areas which allow air to penetrate.

INSOLATION—Technically, this refers to horizontal solar radiation. The term is often used more loosely, however, to refer to all types of solar radiation.

MICROCLIMATE—The climate of a defined local area, such as a house or building site, formed by a unique combination of wind, topography, solar exposure, soil, and vegetation of the site.

PASSIVE SOLAR SYSTEM—A system that relies on natural convection, conduction, and radiation for transfer and storage of heat or coolness. In other words a passive system operates on the natural energy available in the immediate environment.

PATHOGEN—Any agent that causes disease.

PERFORMANCE STANDARDS—Performance standards mandate a building's efficiency by requiring that it be designed to conserve less energy than a predetermined "energy budget" for a similar building in that microclimate. Specific design details are left to the designer's discretion, so long as the final product performs efficiently.

PRESCRIPTIVE STANDARDS—Prescriptive standards mandate that a building's efficiency be met by specific design and construction requirements, i.e., window area, wall construction, insulation thickness, etc.

PLENUM—A chamber in an air-handling system for equalizing or distributing air flow.

RETROFIT—Attaching solar systems or equipment to existing structures, as in the case of a roof-mounted water heater or an add-on greenhouse.

R-VALUE—"R" stands for thermal resistance to winter heat loss or summer heat gain and is more accurate than inches as a means of designating insulation performance. R-19 is the level of insulation recommended by the National Bureau of Standards.

STACK EFFECT—See "Thermal Chimney."

TERTIARY SEWAGE TREATMENT—Although this has no official definition, it is generally accepted that after going through three separate stages of treatment, the resulting water is 95–99 percent free of pollutants, both biological and chemical.

THERMAL CHIMNEY—Heat rising in this tall structure provides a constant suction which may be used to vent the house, bring warm air from collectors, or pull cool air from rock storage. This creates what is called a "stack effect."

THERMAL MASS—The amount of potential heat storage capacity available in a given assembly or system. Water storage tanks, concrete floors, rocks, and masonry are examples of thermal mass.

TITLE 24 OF THE CALIFORNIA ADMINISTRATIVE CODE—
Title 24 establishes energy conservation standards for new residential and non-residential buildings. Details about this standard are available from the Publications Office of the Energy Commission, 1111 Howe Avenue, Sacramento, California 95825.

TROMBE WALL—A concrete, stone or masonry wall that has vents at regular intervals both along the floor and just below the ceiling. The exterior, south-facing side is painted black and covered with glass. Air is warmed between the glass and wall and circulates by convection through the vents. It is named after Dr. Felix Trombe, one of its developers.

WATERWALLS—A passive technique for collecting solar energy. Waterwalls are usually black, water-filled containers exposed to the sun. These collect and store heat, which is used to warm a living space.

BIBLIOGRAPHY

AN INTRODUCTION TO HEAT PUMPS. John A. Summer. Great Britain, Prism Press, 1976. 55 pages.

BIOLOGICAL CONTROL OF WATER POLLUTION. Joachim Tourbier and Robert W. Pierson, Jr., editors. Philadelphia, Pennsylvania, University of Pennsylvania Press, 1976. 340 pages.

BUILD YOUR OWN SOLAR WATER HEATER. Stu Campbell. Charlotte, Vermont, Garden Way Publishing, 1978. 109 pages.

CALIFORNIA SOLAR DATA MANUAL. Paul Berdahl. Sacramento, California Energy Commission, 1978. 318 pages.

THE CITY PEOPLE'S BOOK OF RAISING FOOD. Helga and William Olkowski, Emmaus, Pennsylvania, Rodale Press, Inc., 1975. 228 pages.

COMMUNITY GARDENS IN CALIFORNIA. Rosemary Menninger. Sacramento, State of California, Office of Appropriate Technology, 1977. 39 pages.

DESIGN WITH CLIMATE: Bioclimatic approach to architectural regionalism. Victor Olgyay. Princeton University Press, 1963. 190 pages.

ENERGY, ENVIRONMENT, AND BUILDING. Philip Steadman (Urban and Architectural Studies No. 3). New York, Cambridge University Press, 1975. 287 pages.

ENERGY FOR SURVIVAL: The Alternative to Extinction. Wilson Clark. Garden City, New York, Anchor Press, 1974. 652 pages.

ENERGY PRIMER: Solar, water, wind, and biofuels. Updated and revised edition by Richard Merrill and Thomas Gage. Menlo Park, California, Portola Institute, 1978. 256 pages.

THE FIRST PASSIVE SOLAR CATALOG. David A. Bainbridge. Davis, California, The Passive Solar Institute, 1979. 71 pages.

THE FIRST PASSIVE SOLAR HOME AWARDS. Franklin Research Center. Washington, D.C. United States Department of Housing and Urban Development. Office of Policy Development and Research in cooperation with the United States Department of Energy, 1979. 226 pages.

THE FOOD AND HEAT PRODUCING SOLAR GREENHOUSE: Design, construction, operation. Rick Fisher and Bill Yanda. Santa Fe, New Mexico, John Muir Publications, 1976. 161 pages.

GOODBYE TO THE FLUSH TOILET: Water-saving alternatives to cesspools, septic tanks and sewers. Edited by Carol Hupping Stoner. Emmaus, Pennsylvania, Rodale Press, Inc., 1977. 285 pages.

LOW-COST, ENERGY-EFFICIENT SHELTER FOR THE OWNER BUILDER. Edited by Eugene Eccli. Emmaus, Pennsylvania, Rodale Press, Inc., 1975. 408 pages.

MAN, CLIMATE AND ARCHITECTURE. B. Givonni, New York, Elsevier Publishing Company, Ltd., 1969.

NATURAL SOLAR ARCHITECTURE: A passive primer. David Wright AIA. Van Nostrand Reinhold Company, 1978. 245 pages.

OTHER HOMES AND GARBAGE: Designs for self-sufficient living. Jim Leckie, Gil Masters, Harry Whitehouse, and Lily Young. San Francisco, Sierra Club Books, 1975. 302 pages.

THE PASSIVE SOLAR ENERGY BOOK: A complete guide to passive solar homes, greenhouse and building design. Edward Mazria. Emmaus, Pennsylvania, Rodale Press, 1979. 435 pages.

REGULATIONS ESTABLISHING ENERGY CONSERVATION STANDARDS FOR NEW RESIDENTIAL AND NEW NON-RESIDENTIAL BUILDINGS. Sacramento, California Energy Commission, 1978. Residential—40 pages. Nonresidential—various paging.

REHAB RIGHT: How to rehabilitate your Oakland house without sacrificing architectural assets. Oakland, California. Planning Department. Oakland, California, City of Oakland, 1978. 140 pages.

RESIDENTIAL WATER CONSERVATION. Murray Milne. Davis, California, California Water Resources Center, University of California, 1976. 468 pages.

RESOURCE RECOVERY: Truth & consequences. Marchant Wentworth. Washington, D.C., Environmental Action Foundation, 1977. 77 pages.

SAMPLE WARRANTIES FOR SOLAR ENERGY EQUIPMENT. Sacramento, Solar/Insulation Unit, Department of Consumer Affairs, 1978. 19 pages.

SENSIBLE SLUDGE: A new look at a wasted natural resource. Jerome Goldstein. Emmaus, Pennsylvania, Rodale Press, Inc., 1977. 184 pages.

SOFT ENERGY PATHS: Towards a durable peace. Amory B. Lovins. San Francisco/New York, N.Y., Friends of the Earth International/Harper & Row–Colophon, 1977. 231 pages.

SOLAR CONTROL AND SHADING DEVICES. Aladar Olgyay and Victor Olgyay. Princeton University Press, 1957. 201 pages.

SOLAR DWELLING DESIGN CONCEPTS: A basic guide to solar heating and residential design. New York, Drake Publishers, Inc., 1977. 144 pages.

SOLAR ENERGY: Fundamentals in building design. Bruce Anderson. Harrisville, New Hampshire, Total Environmental Action, Inc., 1977. 374 pages.

SOLAR FINANCING POSSIBILITIES IN CALIFORNIA Spring, 1978. Sacramento, California Energy Commission, 1978. 6 pages.

SOLAR FOR YOUR PRESENT HOME: San Francisco Bay Area edition. Prepared by Berkeley Solar Group. Sacramento, California Energy Commission, 1977. 163 pages.

THE SOLAR GREENHOUSE BOOK. Edited by James McCullagh. Emmaus, Pennsylvania, Rodale Press, Inc., 1978. 328 pages.

THE SOLAR HOME BOOK: Heating, cooling, and designing with the sun. Bruce Anderson. Harrisville, New Hampshire, Cheshire Books, 1976. 297 pages.

SUN ANGLES FOR DESIGN. Robert Bennett. Bala Cynwyd, Pennsylvania, Roberto Bennett, 1978. 77 pages.

SUN/EARTH: How to use solar and climatic energies today. Robert L. Crowther AIA. Crowther/Solar Group, Denver, Colorado. 1977. 232 pages.

THE TOILET PAPERS. Sim Van der Ryn. Santa Barbara, Capra Press, Inc., 1978. 127 pages.

UNDERGROUND DESIGNS. Malcolm Wells. Brewster, Massachusetts, Malcolm Wells, 1977. 87 pages.

VILLAGE HOMES: Solar house designs. David Bainbridge, Judy Corbett, and John Hofacre. Emmaus, Pennsylvania, Rodale Press, 1979. 192 pages.

YOUR ENERGY-EFFICIENT HOUSE: Building and remodeling ideas. Anthony Adams. Charlotte, Vermont, Garden Way Publishing, 1975. 118 pages.

ABOUT FRIENDS OF THE EARTH

Friends of the Earth is an activist environmental lobbying organization. We seek to preserve the natural world not solely for its own sake but to provide an environment hospitable to man.

We are one of the earliest opponents of the dangers posed by nuclear power. We support solar energy and other clean and renewable energy sources. We are leaders in the struggle to preserve the vast Alaskan wilderness as a natural resource for all Americans. We fight in Congress and the courts to implement clean air standards that will make city air breathable. We seek to save whales and other endangered species from extinction.

Thousands of Friends of the Earth members, working together, are making a difference on these issues. We are part of a community of lobbyists, lawyers, and local activists, all supported by a highly informative newspaper and a book publishing program that help to spread the word.

Our greatest resource, however, is people — people who are informed, committed and willing to support action.

Friends of the Earth is headed by David R. Brower, a leading conservationist for more than four decades. He led fights to save the California redwoods, to keep dams out of Dinosaur National Monument and the Grand Canyon, and to preserve natural areas ranging from the Georgia sea islands to Point Reyes in California.

Described by former Interior Secretary Stewart Udall as "the most effective single person on the cutting edge of conservation in this country," Brower was nominated in 1978 for the Nobel Peace Prize for his "persistence in developing the public conscience to the delicate balance of life on our earth."

Brower believes that a conservationist's effectiveness is focussed and multiplied through the power of an organization and that the record of Friends of the Earth is proof of this philosophy.

A RECORD OF COMMITMENT

Nuclear power

Nuclear power is a danger to people and society. David Lilienthal, first head of the Atomic Energy Commission, now warns, "Once a bright hope shared by all mankind, including myself, the rash proliferation of atomic-power plants has become one of the ugliest clouds overhanging America." Friends of the Earth, always one of the most adamant opponents of nuclear power, has exercised decisive influence in several cancellations, suspensions and shutdowns.

Clean energy

True energy independence requires being able to tap renewable energy sources locally that no one can monopolize or interrupt.

The first comprehensive plan for a sustainable energy future was drawn by Amory Lovins, Friends of the Earth's London representative, and published in *Foreign Affairs*. Called the "soft energy path," it is a route to reliance half a century from now solely on renewable energy sources — solar energy and its derivatives, including wind- and water-power, and the conversion of organic matter into fuels. Energy conservation and frugal use of fossil fuels, he proves, will get us through the transition period.

Wildlife

Friends of the Earth believes in natural diversity. We have fought for protection and humane treatment for many animal species, and, as important, have fought to preserve their habitats so that they may survive.

Wildlands

Less than one percent of the American land remains wild. To save this national resource, Friends of the Earth has invested its own resources.

• We successfully campaigned for the Omnibus Parks Bill of 1978, which set up 150 new parks, wilderness areas, wild rivers and new study areas in 44 states — the biggest additions to the park system in American history.

• Friends of the Earth is a co-founder and leader of the Alaska Coalition, which is mounting the greatest conservation lobbying effort in history to induce Congress to set aside many millions of unspoiled acres in Alaska as national parks, wildlife refuges, wild and scenic rivers, and wilderness.

MUCH REMAINS TO BE DONE

A conservation battle is never permanently won. What is once protected must be fought for again and again as new assaults are mounted. Seals, redwoods and whales are still being decimated. Nuclear power plants continue to produce radioactive wastes that no one yet knows how to isolate safely from the human environment for the half-million years during which they remain dangerous. And chemical wastes still contaminate our air, our water and our land.

The issues are tough; the solutions elusive and complex. Yet people working together, as a community, can effectively insist on changes that enhance the quality of their lives. "The moment one definitely commits oneself," W. H. Murray notes, "then Providence moves too. A whole stream of events issues from the decision, which no man could have dreamt would have come his way."

YOU MAKE A DIFFERENCE

We invite your support — because we believe that your commitment will make a difference. For membership application, see last page.

RESOURCES FROM
FRIENDS OF THE EARTH

SUN!
A Handbook For the Solar Decade

Edited by Stephen Lyons

The Sun's rays inspired life on earth and stocked the planet with the fuel we've nearly exhausted. It's time we acknowledge the Sun's importance to our future.

By relying on the energy of the Sun, we can put an end to the costly and dangerous nuclear experiment. The solar future will be cleaner and freer, more equitable and enjoyable than the present.

Sun! is the official book of the International Sun Day movement.

"...This is an overview of the social, economic, and technical advantages of solar energy. There is an extensive bibliography and a listing of organizations active in this field. Recommended..."
—*Library Journal*

"...a crusading, provocative, and prophetic work."
—*Los Angeles Times*

364 pages
$2.95

Frozen Fire:
Where Will It Happen Next?

By Lee Niedringhaus Davis

An LNG accident could be as bad as a reactor meltdown, and the major exporters of liquefied natural gas are OPEC countries.

But the gas industry wants to make it common, building terminals on most American coasts, Europe and Japan, bringing 125,000-cubic-meter shiploads of it in from the Middle East, Indonesia, and Alaska.

US trade in LNG is insignificant now, and world trade is small compared to industry plans. So there is still time to avert the dangers. If we do not, recent horrors with liquefied gases in Spain, Mexico, England, Abu Dhabi, and Staten Island may be the harbingers of a fearful future.

Lee Davis's book is the first major study of LNG prepared for the general reader. It tells everything you should know about the stuff before they try to put it in your town.

"An impressive and powerful compendium of information on the dangers of another chemical threat to the so-called civilized world."
—*John G. Fuller*

298 pages
Paperback: $6.95

The Energy Controversy
Soft Path Questions and Answers

By Amory Lovins and his critics
Edited by Hugh Nash

In the last two years the theories of Amory Lovins have been condemned and defended; criticism has been led by advocates of the big, centralized energy systems that are now leading us into one energy crisis after another. The debate climaxed in joint hearings of two Senate committees, where Lovins refuted his major critics, point by point.

The Energy Controversy recaps those hearings, presenting Lovins and his opponents in dramatic dialogues on this most central issue of our time. How much does the soft path cost? What dangers lurk on our present hard road? How much would *it* cost? And how do we shift from one to the other?

Lovins is "...one of the Western world's most influential energy thinkers. His visionary essay, 'Energy Strategy: The Road Not Taken?', is described by population biologist Paul Ehrlich as 'the most influential single work on energy policy written in the last five years.' 'He has put a lot of things together that other people have been dealing with in pieces for years,' says John Holdren, an energy expert at the University of California at Berkeley. 'Because of him, the energy debate will never be the same.' Lovins's thesis has yet to be disproved, and his influence is widening daily. . . .The day after he talked with Lovins, President Carter told an international energy conference that the world should consider alternatives to nuclear power. . . ."
—*Newsweek*

450 pages
Cloth: $12.50
Paperback: $6.95

Soft Energy Paths
Toward a Durable Peace

By Amory B. Lovins, introduction by Barbara Ward

Soft Energy Paths crowns the critical work of Amory Lovins and FOE with a new program for energy sanity..This is an epoch-making book, the most important Friends of the Earth has published.

Lovins's path can take us around nuclear power, free industrial nations of dependence on unreliable sources of oil and the need to seek it in fragile environments, and at the same time enable modern societies to grow without damaging the earth or making their people less free.

Soft Energy Paths provides a conceptual and technical basis for more efficient energy use, the application of appropriate alternative technologies, and the clean and careful use of fossil fuels while soft technologies are put in place.

Harper & Row/Colophon edition
231 pages
$3.95

Progress As If Survival Mattered

Edited by Hugh Nash

The environmental book of 1978, *Progress As If Survival Mattered* presents Friends of the Earth's comprehensive and realistic program to take America away from energy crises and resource shortages. . . and charts the way to an abundant healthy and free future. This "Handbook for a Conserver Society" includes the thinking of our finest writers on energy, environment, and society.

"Here is a book undoubtedly headed for fame. . . an important anthology of long articles on issues which are critical. . .a guidebook for citizen action . . .an all-around introduction to the ills and successes of our planet, it should find its way into the homes of almost anyone who reads more than the Sunday comics. . . .Highly recommended."
—*Library Journal*

320 pages
$6.95

Pathway to Energy Sufficiency
The 2050 Study

By John Steinhart

Pathway to Energy Sufficiency is a hardheaded, hopeful look at an American society seventy years in the future. The authors draw a picture of what it would be like to live along a soft energy path, using 36 percent of the energy we used in 1975. They tell how this would affect our houses, travel, countryside, factories, gardens, and public institutions. They find this low energy society decentralized, more rural, less hurried, and healthier.

"The benefits that will accrue to a society that accepts the low-energy scenario could be most attractive. If people want a world in which people restrain their numbers and appetites, people can achieve it — and will prefer it to the grim alternatives."

David R. Brower

96 pages, illustrated
$4.95

Reel Change
A Guide to Films on Appropriate Technology

by Soft-Aware Associates, Inc.

The creation of renewable energy, agriculture, and social alternatives is being helped — and hampered — by the films available about alternative technology.

Reel Change is a guide to the best, and the worst, of rental films, from inspirational documentaries to "how-to" films and science lessons. Some are highly recommended, some aren't — it depends on how useful they are. The authors give readily accessible ratings to more than 80 movies and provide a directory of distributors.

56 pages, paperback
$3.50

The Whaling Question

The Inquiry of Sir Sydney Frost

This is the book that made the Australian government end Australia's industry, close its waters to whaling, and become a leader in the international movement to stop commercial whaling. The Whaling Question is a verbatim copy of the Inquiry undertaken by the Australian government to consider all the data available about whaling from all over the world. The reference needed by whale-loving activists.

340 pages, paperback
black and white photographs
$6.95

The Whale Manual

By Friends of the Earth Staff

The Whale Manual lays out the latest facts and figures about the great whales. Population estimates, habitats, and how quickly they are being killed, by whom, for what, and how — and what could be used as alternatives to whale products.

A special section outlines Friends of the Earth's controversial program to preserve the endangered Bowhead, with the help of the Eskimos who hunt it.

The Whale Manual is an invaluable source for readers committed to saving our planet's largest creatures.

168 pages
$4.95

The Energy and Environment Bibliography

Prepared by Betty Warren

A unique guide to information on energy. The Energy and Environment Bibliography lists hundreds of books, magazines, newsletters, films, and activist groups, with annotations on each.

Just revised, the Bibliography is the most up-to-date guide to resource material for everything from arming the reader to fight a nuclear plant to building a roof-top solar collector. Published with a grant from Friends of the Earth Foundation.

100 pages, 8½ × 11
$3.50

To: Friends of the Earth
124 Spear Street
San Francisco, CA 94105

Please Join Us.

☐ Please enroll me for one year in the category checked, entitling me to *Not Man Apart* and discounts on selected FOE books.
(Contributions to FOE are not tax-deductible.)

☐ Regular = $25   ☐ Spouse = add $5
☐ Supporting = $35*   ☐ Life = $1000***
☐ Contributing = $60** ☐ Patron = $5000***
☐ Sponsor = $100**   ☐ Retired = $12
☐ Sustaining = $250**   ☐ Student/Low Income = $12

*Will receive free a paperback volume from our *Celebrating the Earth* Series.
**Will receive free a volume from our *Earth's Wild Places* Series.
***Will receive free a copy of *Headlands* (our award-winning, gallery-format book).
☐ Check if you do not wish to receive your bonus book.

☐ Please accept by *deductible* contribution of $ _____ to Friends of the Earth Foundation *(checks must be made to FOE Foundation)*.

Please send me the following FOE Books:

Number	Title, price (members' price)	Cost
_____	*Energy Controversy* at $12.50 ($10.00)	_____
_____	paperback *Energy Controversy* at $6.95 ($5.95)	_____
_____	*Frozen Fire* at $6.95 ($5.95)	_____
_____	*SUN! A Handbook for the Solar Decade* at $2.95 ($2.25)	_____
_____	*Soft Energy Paths*, H&R ed. at $3.95 ($3.25)	_____
_____	*Reel Change* at $3.50 ($3.00)	_____
_____	*Progress As If Survival Mattered* at $6.95 ($5.75)	_____
_____	*Pathway to Energy Sufficiency* at $4.95 ($3.95)	_____
_____	*The Whaling Question* at $6.95 ($5.95)	_____
_____	*The Whale Manual* at $4.95 ($3.95)	_____
	Other FOE Titles:	

_____   _____
_____   _____
_____   _____

Subtotal _____
6% tax on Calif. delivery _____
Plus 5% for shipping/handling _____

☐ Send full FOE Books catalogue.   TOTAL _____
☐ VISA   ☐ Mastercharge
Number _____ Expiration date _____
Signature _____
Name _____
Address _____
City _____ State _____ Zip _____